T0094014

Teaching Data Analytics

Data Analytics Applications

Series Editor:
Jay Liebowitz

PUBLISHED

Big Data in the Arts and Humanities: Theory and Practice
by Giovanni Schiuma and Daniela Carlucci
ISBN: 978-1-4987-6585-5

Data Analytics Applications in Education
by Jan Vanthienen and Kristoff De Witte
ISBN: 978-1-4987-6927-3

Data Analytics Applications in Latin America and Emerging Economies
by Eduardo Rodriguez
ISBN: 978-1-4987-6276-2

Data Analytics for Smart Cities
by Amir Alavi and William G. Buttlar
ISBN: 978-1-138-30877-0

Data-Driven Law: Data Analytics and the New Legal Services
by Edward J. Walters
ISBN: 978-1-4987-6665-4

Intuition, Trust, and Analytics
by Jay Liebowitz, Joanna Paliszkiewicz, and Jerzy Gołuchowski
ISBN: 978-1-138-71912-5

Research Analytics: Boosting University Productivity and Competitiveness through Scientometrics
by Francisco J. Cantú-Ortiz
ISBN: 978-1-4987-6126-0

Sport Business Analytics: Using Data to Increase Revenue and Improve Operational Efficiency
by C. Keith Harrison and Scott Bukstein
ISBN: 978-1-4987-8542-6

https://www.crcpress.com/Data-Analytics-Applications/book-series/CRCDATANAAPP

Teaching Data Analytics
Pedagogy and Program Design

Edited by
Susan A. Vowels
Katherine Leaming Goldberg

CRC Press is an imprint of the
Taylor & Francis Group, an informa business
AN AUERBACH BOOK

CRC Press
Taylor & Francis Group
6000 Broken Sound Parkway NW, Suite 300
Boca Raton, FL 33487-2742

Printed on acid-free paper

International Standard Book Number-13: 978-1-138-74414-1 (Hardback)

Visit the Taylor & Francis Web site at
http://www.taylorandfrancis.com

and the CRC Press Web site at
http://www.crcpress.com

Susan would like to dedicate this book to the memory of her father, Chief Master Sergeant Milton R. Vowels, U.S.A.F.

Kate would like to dedicate this book to her husband, Mark Goldberg, her parents, Judi and Spicer Leaming, and her sons, Matt and Josh Goldberg.

Contents

Preface: Teaching Data Analytics—A Primer for Higher Education

Introduction

Thomas Davenport's 2006 article, "Competing on Analytics," brought into focus a trend that at that time was well underway but not well understood. He recognized that successful firms were already seeing the effective use of analytics as a powerful tool in developing and maintaining a competitive advantage. Since then, the use of analytics has exploded, becoming one of the most powerful instruments for organizational success in a multitude of scenarios, not just business. The need for analytics skills has, in turn, led to a burgeoning growth in the number of analytics and decision science programs in higher education developed to feed the need for capable employees in this area.

The very size and continuing growth of this need means that there is still space for new program development; indeed, in examining the gap between business analytics offerings by business schools and corresponding industry needs, Turel and Kapoor (2016) found that there was a significant gap in the maturity level of business analytics offerings. They recommended that those schools wishing to pursue business analytics programs intentionally assess the maturity level of their programs and take steps to close the gap. Katherine Goldberg and I intend this text to inform faculty and administrators who are thinking about adding or enhancing analytics offerings at their institutions by providing a look into best practices from the perspectives of faculty and practitioners. While the importance of this area of study may be considered unquestioned, it is useful to review the connection of data analytics to organizational success, to consider the position of analytics and decision science programs in higher education, and to review the critical connection between this area of study and career opportunities. Key differentiators of this text from others lie in the variety of perspectives gathered, ranging from the scholarly theoretical to the practitioner applied; the breadth of skills discussed, from closely technology-focused to robustly soft human connection skills; and the inclusion throughout of resources for existing faculty to acquire and maintain additional analytics-relevant skills that can enrich their current course offerings.

Data Analytics vs. Data Science

An important concept to be understood by institutions wishing to add or further develop analytics programs is the dichotomy between data analytics and data science. Aasheim, Williams, Rutner, and Gardiner (2015) conducted a study of undergraduate programs at 13 universities offering majors in these areas to determine how institutions distinguished between data analytics and data science. They found that while both types of programs included visualization, data mining, big data, and analytics/modeling, data analytics programs tended to be housed in business schools, including information system programs or other quantitative business programs, whereas data science programs tended to be housed in computer science departments or were interdisciplinary. Turel and Kapoor (2016) found a similar pattern in their study of what they termed "business analytics-related" course offerings from both ranked and unranked business schools. Their findings were that business analytics courses were generally listed within the structure of the business schools, while courses that pertained to data science were more typically offered in interdisciplinary settings and more technically focused than the business analytics programs housed solely within business schools.

Aasheim et al. (2015) also drilled down to the course levels to determine the percentage of universities with programs containing specific types of courses. They found the following topics to be the most prevalent (present in at least 50% of the programs) among both data analytics and data sciences:

- Mathematics/statistics/probability
- Visualization techniques
- Data mining techniques
- Other modeling/analytics techniques (p. 109)

Some topics were more heavily prevalent in either data analytics or data science. At least 50% of data analytics programs contained courses addressing decision-making skills and courses evaluating the applicability of specific tools, which in some cases considered the alignment of such tools to business strategy. In contrast, decision science programs were heavier in courses pertaining to programming, understanding of big data and unstructured data, and data preparation and quality.

Two results of this study were striking. First, neither data analytics nor data science programs in the study contained courses devoted to solely ethical considerations. However, the authors noted that it was possible that ethical considerations were incorporated within other topics; in addition, this study was restricted to a relatively small pool of programs. Of equal significance, study data indicated while courses in decision-making skills were prevalent in data analytics programs, none of the decision science programs contained such courses.

It is important to note that analytics is different in nature than many other types of data discovery in that typically analytics are used to enable organizations to identify a call to action of some sort. These calls to action are not restricted to commercial contexts; indeed, vintage scenarios included in Tableau's "The 5 Most Influential Visualizations of All Time" (n.d.) address such calls to action as the need for municipal sewer systems to prevent cholera epidemics, the vital importance of hospitals and sanitation in preventing deaths in wartime, and a reminder to the French populace of the terrible consequences of Napoleonic nationalism. In their discussion of data analytics in the context of auditing, Braun, Struthers-Kennedy, and Wishna (2017) emphasize that "insight from analytics are the result of the intersection between business awareness and the application of analytics tools and methodologies" (p. 42). This argues that both decision-making and ethics are crucial elements to be included in data analytics programs and perhaps data science programs as well.

It is also essential to understand the importance of human participation in the data analytics process. We can see this in the research framework for the creation of strategic business value from the deployment of big data analytics (BDA) proposed by Grover, Chiang, Liang, and Zhang (2018). Defining BDA as "the application of statistical, processing, and analytics techniques to big data for advancing business" (p. 390), they argue that within the context of business, BDA is an essential tool for the creation of competitive advantage and further that it is unparalleled when looking at recent business trends in its ability to transform existing information technology investments. While their focus is within the context of business, their observations can be generalized to many kinds of organizational settings, including scientific and not-for-profit. With our focus on education, it is relevant to consider how human capital fits within their analysis; and indeed, the human factor is mentioned several times within their study. First, they note that while BDA implementation begins with an information technology orientation, leveraging benefits from it comes from an orientation to the organization's strategy (Grover et al., 2018). Business strategy is developed from human capital, making the optimum use of BDA inextricably linked to effective use of it by various people within the firm. Indeed, Grover et al. (2018) note that some organizations find their greatest challenges in building BDA capability lie in building the human and organizational infrastructure linking data scientists and big data professionals to those in the firm responsible for operations and strategy. This challenge can come from a lack or insufficiency of BDA skills on the part of current employees as well as from difficulty in filling open positions for data scientists and data professionals. Further, it is crucial for organizational leadership to foster corporate culture surrounding the effective use of BDA, enjoining teams to approach problems and challenges from the point of view of data and evidence.

Hence, the focus of this text as a primer for those exploring the addition of data analytics to their curriculum, whether as exercises to be added to existing classes or the addition of individual classes or programs. In this text, the emphasis will be on data analytics rather than on data science, although there are times when we do move a bit into the data science realm.

Organization of the Book

We set the stage by starting with industry perspectives. ADVIZOR Solutions provides business analytics software as well as consulting services. Doug Cogswell, ADVIZOR CEO, and colleagues Eric Cho and Mateo Molina Cordero provide a window into the applied world of data analytics, covering necessary skills and applications. A key element in understanding how to approach data analytics is the concept of developing a model. This can be accomplished using a variety of software tools combined with statistical knowledge. Doug Cogswell and his colleagues begin our discussion by pointing out that while mathematics is essential for data analytics, there is much more involved in successfully reaping benefits from analytics.

Staupell Analytics Group provides targeted technical training to support professionals in the field of fundraising. Marianne M. Pelletier, managing director of Staupell, describes an introductory two-day course that contributes key lessons learned for teaching analytics concepts in higher education venues. She shows us how practitioners apply data analytics in their designated fields, in this case, fundraising for educational institutions, and brings in the important concept of audience when developing compelling visualizations.

In the last chapter of the first section, Dr. Kathryn S. Berkow, who teaches in the Department of Accounting and Management Information Systems at the University of Delaware, describes how she integrates essential professional skills into her analytics course to prepare students for the professional environment they will find when they enter the work world. The best technical analysis still needs an advocate who can provide compelling arguments for the calls to action resulting from the analysis. Dr. Berkow's background in industry provides her with a perspective that spans both the practitioner's world and the academic setting of undergraduate education.

The rest of the book examines pedagogical and program design approaches in data analytics education, beginning with four chapters that outline curricular and cocurricular projects. Dr. Matt North of the Information Systems and Technology Department at Utah Valley University guides the reader through an understanding of the application and differing roles of formative and summative assessment strategies and then applies them to student learning of association rules. He provides us with the mathematical underpinnings of association rules and then discusses how to use R algorithms in RStudio to generate association rules.

Dr. Virginia Miori is chair of the Department of Decision and Systems Sciences at St. Joseph's University. Chapter 5 extends the discussion of the importance of modeling within the field of data analytics; she illustrates how to teach computer simulation, a form of prescriptive analytics using ExtendSim software. As Dr. Miori points out, simulation is an underrepresented data analytics instrument. Her chapter presents directions for student exercises, ties the simulation to a business problem, and points out the statistical tools connected to the exercise, addressing program design and connection to the mission of the institution.

Dr. Stephen Penn of the Business Analytics Department at Harrisburg University of Science and Technology applies the use of games to engender student interest and engagement and so improving student learning of analytical techniques. He offers several examples, in one instance providing directions for a game that can be played by the entire class using Excel and in another describing a simulation game using the Python programming language.

This section closes with a description of a cocurricular experience provided by Yelena Bytenskaya, Elena Gortcheva, and Katherine Goldberg, all affiliated with the University of Maryland University College Master of Data Analytics program. They describe the experience Prof. Bytenskaya and Dr. Gortcheva had with student participation in IBM Watson competitions. Competitions allow student learning to extend beyond the classroom and this cocurricular activity uses an external competition to ground student knowledge of analytics skills while connecting their technical skills with the ability to develop a narrative and a call to action.

The last section in our book offers ideas for program design tactics. A discussion of the competencies needed for the design, implementation, and adoption of the analytics process is brought to us by Dr. Eduardo Rodriguez, Sentry Endowed Chair in Business Analytics at the University of Wisconsin-Stevens Point; Dr. John S. Edwards, Professor of Knowledge Management at Aston Business School; and Dr. Germán Ramírez, President of GRG Education. This chapter can be used to inform educators contemplating the development of student learning outcomes for new or existing programs in data analytics.

This is followed by two implementation models: one based on a single undergraduate course and the other describing a ranked graduate program. Katherine Goldberg of the Department of Business Management at Washington College describes a business analytics course that could be used as a first step in offering an undergraduate data analytics curriculum. The reader will find a treasure trove of material, including a course outline and resources for the software used in the course such as SAP Predictive Analytics, Tableau, and BigML, a machine-learning data analytics platform. Example assignments are also provided. This course became a key component of a new minor in Data Analytics at Washington College that will be launched in Fall 2019.

The last chapter in the text brings together many of the themes reflected throughout. Authored by Dr. Virginia Miori, Dr. Nicolle Clements, and Dr. Kathleen Campbell-Garwood, all of the Decision and System Sciences Department at St. Joseph's University, this chapter describes the development and continuous renewal of St. Joseph's Master of Science in Business Intelligence and Analytics (MSBIA). This mature, ranked data analytics program was designed in such a fashion as to remain relevant regardless of changes in software tools and technology over time.

Conclusion

Of course, with a topic as vast as data analytics pedagogy and program design in higher education, we could not hope to cover everything. However, we do hope that the reader will use this text as a launching point for discussions about how to connect industry's need for skilled data analysts to higher education's need to design a rigorous curriculum that promotes student critical thinking, communication, and ethical skills; for adding new elements to existing data analytics courses; and for taking the next step in adding data analytics offerings, whether it be incorporating additional analytics assignments into existing courses, offering one course designed for undergraduates, or an integrated program designed for graduate students.

Susan A. Vowels, DBA
Washington College

References

Aasheim, C. L., Williams, S., Rutner, P., & Gardiner, A. (2015). Data analytics vs. data science: A study of similarities and differences in undergraduate programs based on course descriptions. *Journal of Information Systems Education, 26*(2), 103–114.

Braun, G., Struthers-Kennedy, A., & Wishna, G. (2017). Building a data analytics program. *Internal Auditor, 74*(4), 41–45.

Davenport, T. H. (2006). Competing on analytics. Competing on analytics. *Harvard Business Review, 84*(1), 98–107.

Grover, V., Chiang, R. H. L., Liang, T., & Zhang, D. (2018). Creating strategic business value from big data analytics: A research framework. *Journal of Management Information Systems, 35*(2), 388–423. doi:10.1080/07421222.2018.1451951.

Tableau (n.d.). The 5 Most Influential Visualizations of All Time. Retrieved from https://www.tableau.com/learn/webinars/5-most-influential-visualizations-all-time

Turel, O., & Kapoor, B. (2016). A business analytics maturity perspective on the gap between business schools and presumed industry needs. *Communications of the Association for Information Systems, 39*, 96–109.

Acknowledgments

We would like to thank Dr. Jay Liebowitz for presenting us with the opportunity and challenge of editing a book about this topic and providing us with invaluable guidance and mentorship. We are also grateful to our publishing editor John Wyzalek and his colleagues at CRC Press/Taylor & Francis Group for their support and patience. Our special thanks go to the authors who contributed chapters to this book for their mastery of various topics pertaining to teaching data analytics and for their ability to share their expert knowledge in an adroit, approachable manner.

This project could not have been completed without the support of our families and colleagues. Susan would like to thank her mother, Dorothy Ann Vowels, and her sister, Karen Vowels Earle, for always believing in her. Katherine is grateful for her husband, Mark, for his patience and encouragement. We are both grateful for our colleagues at Washington College who provided constant moral support.

Susan A. Vowels, DBA
Constance F. and Carl W. Ferris Associate Professor and Chair
Department of Business Management, Washington College
Chestertown, Maryland

Katherine Leaming Goldberg, MS
Director of Advancement Services, Washington College
Chestertown, Maryland

Editors

Dr. Susan A. Vowels is the Constance F. and Carl W. Ferris associate professor and Chair of the Department of Business Management at Washington College. In 2002, she launched the business program's management information systems curriculum, subsequently developing courses in management information systems, enterprise resource planning systems, business intelligence, and business analytics. She is a cofounder and codirector of the College's Information Systems Minor and provided the impetus for Washington College to join the SAP University Alliance, serving as faculty coordinator since 2003. Most recently, along with Dr. Austin Lobo, she co-founded the Data Analytics Minor at Washington College and will serve as program director beginning Fall 2019. Her broad research interests have included attention to how technology can serve strategic corporate goals, methods for infusing ethics into management information systems pedagogy, and how the examination of end user well-being can help us better understand ethical considerations in information systems implementations. Her broad research interests are informed by a liberal arts degree from St. John's College in Annapolis, Maryland; an MBA with a specialization in international business from the University of Delaware; and a DBA from Wilmington University as well as a successful career in industry. Her service in industry included stints as a systems engineer with IBM; as a programmer for firms supporting the advertising and travel industries; and as a consultant supporting an SAP implementation, along with a successful career with a wholesale millwork firm, moving from functional responsibility for information systems to serving as vice president and a member of the board of directors. She sees data analytics as a unifier of disparate information and an essential tool for organizational decision-making.

Katherine Leaming Goldberg serves two roles at Washington College in Chestertown, Maryland. She is the Lecturer of Business Analytics in the Department of Business Management at Washington College. She is also the Director of Advancement Services, helping the college fundraise by applying data analytic approaches. Her research interests include methods of using predictive text and cluster analytics to increase fundraising at nonprofit organizations. Her love for data started with her undergraduate degree when she designed her own major in Mathematical Biology at Randolph-Macon Woman's College in Lynchburg, Virginia. She holds a certificate in Fundraising Operations from the Susan B. Glasscock School of Continuing Studies at Rice University. She completed her Masters of Data Analytics from University of Maryland University College (UMUC) and was inducted as a member of Phi Kappa Phi. She is an instructor in the Fundraising Operations program at Rice University, where she recently created an on-demand course to teach nonprofit professionals about gift processing. She is also a Teaching Assistant in the Masters of Data Analytics program at UMUC.

Contributors

Dr. Kathryn S. Berkow is an Assistant Professor in the Department of Accounting & Management Information Systems at the University of Delaware. Following undergraduate degrees from the University of Delaware, she earned a PhD in Applied Mathematics and Statistics from Stony Brook University. Kathryn recently returned to UD from a career in financial services analytics. Since then, her focus has been on developing engaging courses in business analytics—on campus and online, undergraduate and graduate—and continuing research in global equity and fixed income markets.

Yelena Bytenskaya received a bachelor degree in computer science from the University of Maryland University College (UMUC) in the spring of 2002 and graduated with Summa Cum Laude honors. In the fall of 2008, she received a Master's degree in Information Technology Database specialization also from UMUC. After graduation, Yelena decided that it was her turn to help new students in a program. She was hired as a graduate teaching assistant, and it has been her second job since then. Two years ago, she was offered an opportunity to coach a team of students for the Watson Analytics competition. It was an excellent experience and professional growth. She strongly believes that "A" students are not born. They are made through daily hard work.

Dr. Kathleen Campbell-Garwood is an Assistant Professor in the Department of Decision & System Sciences and has been a member of the faculty at St. Joseph's University since 2004. Her teaching primarily focuses in data mining and modeling with the goal of introducing real data and analytical techniques to students. Her research interests include data visualization (best practices and techniques), rank order comparisons (with a focus on sustainability rankings), modeling applications with real-world settings (including identifying the most impoverished for the Fe y Alegria:Bolivia), and data collection and analysis related to science, technology, engineering, and mathematics (STEM). Recently, she has combined these interests working with the United Nations' Principles for Responsible Management Education (PRME) group to collect, organize, and visualize through interactive dashboards the efforts being made in management schools to address the UN sustainable development goals. In the venture, she works to oversee and educate undergraduate and graduate students in data collection and visualization.

Eric Cho is a business consultant at ADVIZOR Solutions, a Business Intelligence software company. As part of the consulting team at ADVIZOR Solutions, he has enabled clients to take charge of their data and utilize it to produce valuable insights, create coachable performance metrics, and enhance collaboration across teams. For the past 4 years and across two dozen clients, he has designed and deployed dashboards, created models, provided strategic assessments, and conducted training and workshops on predictive modeling. Eric holds a degree in mathematics and physics from the University of Chicago. In his free time, he enjoys analyzing board games, biking along Lake Michigan, snowboarding, camping, and cooking.

Dr. Nicolle T. Clements is a PhD statistician and an Assistant Professor in the Department of Decision System Sciences at Saint Joseph's University, where she also serves as the Academic Coordinator of the MSBIA program. Dr. Clements holds a doctorate in statistics from Temple University's Fox School of Business, a Master of Science in statistics from Virginia Polytechnic Institute and State

University (Virginia Tech), and a Bachelor of Science in mathematics from Millersville University. Her PhD research was in the area of high-dimensional multiple testing procedures, and she currently conducts applied work in spatial and environmental applications of analytics, multiplicative time-series modeling, and statistical analysis of substance abuse treatments. Much of her research is focused on adapting standard statistical models to be used in nontraditional applications. She is currently working on a grant with Dr. Virginia Miori through the Pennsylvania Department of Agriculture to collect and analyze data in order to evaluate the PA Preferred brand. A full list of Dr. Clements' scholarly publications can be found here: https://www.sju.edu/about-sju/faculty-staff/nicolle-clements-phd. In addition to her research, Dr. Clements frequently teaches courses at the undergraduate, graduate, and executive level on topics such as statistics, data mining, and R statistical programming.

Doug Cogswell is the founder and current President and CEO of ADVIZOR Solutions, a Business Intelligence software company that is all about enabling people to better understand and analyze their data. Under Doug's leadership, ADVIZOR has been delivering data discovery solutions in a variety of industries for over 12 years. By combining cutting-edge software with the services of a team of data and analytics experts, ADVIZOR gives people quick, easy access to their data in a visual, interactive format that is transforming the way they make decisions, improving overall performance, and creating a culture of analytics. With a degree in physics and engineering from Dartmouth, an MBA from Harvard, strategy consulting experience with both Bain and Booze Allen, and over 15 years in the BI sector, Doug has extensive data analytics and client strategy expertise. He is a thought leader in the world of data discovery and analysis. Doug has participated on the Boards of the Chicagoland Chamber of Commerce, the Information Technology Association of Illinois, and is a frequent speaker at national and regional conferences. He is also heavily involved in his church. And when he is not working, he enjoys hockey, skiing, running, backpacking, and pretty much anything outdoors.

Dr. John S. Edwards is Emeritus Professor and Professor of Knowledge Management at Aston Business School, Birmingham, UK. He holds MA and PhD degrees from Cambridge University. His interest has always been in how people can and do (or do not) use models and systems to help them do things. At present his principal research interests include how knowledge affects risk management, investigating knowledge management strategy and its implementation, and the synergy between knowledge management, analytics, and big data. He has written more than 75 peer-reviewed research papers and three books on these topics. He is consulting editor of the journal *Knowledge Management Research & Practice.*

Dr. Elena Gortcheva is Program Chair for the Master of Science in Data Analytics (MSDA) in the Graduate School at the University of Maryland University College (UMUC). Her professional interests focus on machine learning and intelligent systems, and she teaches courses in machine learning and big data analytics. Dr. Gortcheva joined UMUC in 2003 after a 15-year computer engineering career in academia as a professor/researcher and industry consultant. She holds a Master of Science in electronic engineering and earned her PhD in computer engineering from the Bulgarian Academy of Sciences. The results of her research are published in numerous articles, presented at several scientific conferences, and have led to different commercialization projects. Dr. Gortcheva implemented a wide array of advanced technologies and innovative approaches to support and enrich the student's learning experience including through course development and industry-oriented research projects, Oracle Labs, Virtual Lab, and open source software such as R, Python, Hadoop, and Spark. She has implemented her professional experience to the curricula development for Master of Science in Data Analytics and Database System Technology.

Dr. Virginia M. Miori is a Full Professor at Saint Joseph's University in Philadelphia, Pennsylvania. She is a graduate of the LeBow College of Business at Drexel University, holding a doctoral degree in Operations Research. She also holds an MS in Operations Research from Case Western Reserve University and an MS in Systems

Engineering and Transportation from the University of Pennsylvania. Dr. Miori has participated in an Ignatian Pedagogy seminar along with an Ignatian Leadership Program at Saint Joseph's University. Dr. Miori is the current chair of the Department of Decision and System Sciences. She was the Academic Coordinator of the Master of Science in Business Intelligence and Analytics at Saint Joseph's University for 8 years, an internationally ranked program. She has 18 years of teaching experience and over 12 years of industry experience in developing and implementing statistical and operations research models in the area of supply chain/logistics. She is active in research in the areas of supply chain, scheduling, and predictive analytics and has received several research awards.

Mateo Molina Cordero leads the services team at ADVIZOR Solutions, Inc. The team works with business users, IT, and executives to customize Business Intelligence solutions. They identify key questions and data available; create metrics and statistical models; and deliver dashboards, training, and strategic advice. Mateo is a data specialist and consultant focused on helping organizations develop a culture of analytics and business intelligence. His priority is to deliver practical solutions to create value, improve efficiency, and eliminate operational bottlenecks.

Dr. Matt North is a Professor in the Information Systems and Technology department at Utah Valley University. His teaching expertise is in database design, development and administration, data analytics, and geographic information systems (GIS). Research areas include data analytics, GIS, and technology pedagogy. He is the author of two books, *Life Lessons & Leadership* and *Data Mining for the Masses*, as well as numerous journal papers, articles, book chapters, and conference presentations. Prior to his work in academia, he was a software engineer and risk analyst at eBay, and consulted internationally. An award-winning professor and scholar, he is a Fulbright alumnus, and the recipient of the Ben Bauman Award for Excellence, the Gamma Sigma Alpha Outstanding Professor Award, and the 2018 UVU Alumni Outstanding Educator Award.

Marianne M. Pelletier has 30 years' experience in the nonprofit sector, primarily in prospect research and analytics. An early adopter of statistical analysis for fundraising, Pelletier has provided research, annual giving management, and analytics for Harvard, Lesley, Southern New Hampshire, Carnegie Mellon, and Cornell Universities. She now runs the Staupell Analytics Group, an all-about-data firm. She is a recipient of the Anne Castle Award, a lifetime achievement award from the New England Development Research Association, and has served on several boards relating to prospect research and analytics in fundraising. Pelletier has a BA from Rockford University and an MBA from Southern New Hampshire University. Her book, *Building Your Analytics Shop: A Workbook for Nonprofits,* is available on Amazon and on her company's Web page, www.staupell.com. Her online course, "The Analytics Journey Throughout Your Campaign," is also available online at https://staupell-analytics-group-online-workshops.teachable.com/p/campaign-analytics-an-overview1.

Dr. Stephen Penn earned his Doctor of Management from University of Maryland University College. His dissertation focused on data-driven decision-making. He earned an MBA from Frostburg State University and bachelor's degrees in mathematics and Russian from University of Texas at Arlington. He is certified in project management. Stephen Penn has worked in information technology for more than 20 years, specializing in database development and analytics. His experience in analytics includes projects focused on student achievement in higher education, insurance fraud, and workforce optimization. He is currently an Associate Professor of Business Analytics at Harrisburg University of Science and Technology in Harrisburg, Pennsylvania.

Dr. Germán A. Ramírez leads the higher education consulting practice at GRG Education, aimed at assisting universities in matters of strategy, organizational effectiveness, digitalization, and internationalization. Prior to this, he worked for 18 years at Laureate International Universities, where he held, among others, the following executive positions: Chief Academic Officer for Europe, President of Network Products, President Universidad de Las Americas Chile, and President of Laureate Central America. Ramirez holds an EdD from Harvard University. He also holds an MPA from the National Institute of Public Administration (Spain), a law degree from Los Andes University (Colombia) and a BBA from CESA Business School (Colombia). He has been a university lecturer in the fields of constitutional law, economics, and business administration.

Dr. Eduardo Rodriguez is the Sentry Endowed Chair in Business Analytics University of Wisconsin-Stevens Point. In his work, he has created the Analytics Stream of the MBA at the University of Fredericton; Analytics Adjunct Professor at Telfer School of Management at Ottawa University; Corporate Faculty of the MSc in Analytics at Harrisburg University of Science and Technology, Pennsylvania; Senior Associate Faculty of the Center for Dynamic Leadership Models in Global Business at The Leadership Alliance, Inc., Toronto Canada; and Principal at IQAnalytics, Inc. Research Centre and Consulting Firm in Ottawa Canada. He has been Visiting Scholar at Chongqing University, China, and EAFIT University for the Master of Risk Management. Eduardo has extensive experience in analytics, knowledge, and risk management mainly in the insurance and banking industry. He has been the Knowledge Management Advisor and Quantitative Analyst at EDC Export Development Canada in Ottawa; Regional Director of PRMIA (Professional Risk Managers International Association) in Ottawa; and Vice-President Marketing and Planning for Insurance Companies and Banks in Colombia. Moreover, he has worked as part-time professor at Andes University, a CESA in Colombia, was an author of six books in analytics and reviewer of several journals, and has publications in peer-reviewed journals and conferences. He created and Chaired the Analytics Think-Tank, organized and Chaired the International Conference in Analytics ICAS, is a member of academic committees for conferences in Knowledge Management, and is an international lecturer in the analytics field. Eduardo holds a PhD from Aston Business School, Aston University in the UK, an MSc Mathematics from Concordia University, Montreal, Canada, a Certification of the Advanced Management Program from McGill University, Canada, and an MBA and Bachelor in Mathematics from Los Andes University, Colombia. His main research interest is in the field of Analytics and Knowledge Management applied to Enterprise Risk Management.

SECTION I
Industry Perspective

1

IT'S NOT ALL ABOUT THE MATH*

DOUG COGSWELL, ERIC CHO,
AND MATEO MOLINA CORDERO

ADVIZOR Solutions, Inc.

Contents

This chapter guides a user through the process of building effective predictive models. It covers nine general concepts:

- What Is a Predictive Model?
- The Business Question
- Linear Regression vs. Classification
- Base Population
- Target Field
- Explanatory Factors
- Thinking Through Factors
- Model Training and Output
- Eight Step Process

How to Read

Supplementing the main body of text and scattered throughout the workbook are several types of textboxes.

Doug's Helpful Tips—These are suggestions, thoughts, or comments from ADVIZOR CEO Doug.
Mateo's Main Point—These are summaries of each section from ADVIZOR's lead business consultant Mateo.
Linear Regression Case Study—The application of the section's topic in a case study that predicts a target that is a numeric value (linear regression).
Classification Regression Case Study—The application of the section's topic in a case study that predicts a target that classifies data into two categories (classification).

What Is a Predictive Model?

A predictive model is a mathematical and statistical description of patterns in a set of data. The model then applies to a new set of data to make predictions. Here are some examples:

- Using data on past accidents to calculate the risk of accidents for policy holders in car insurance.
- Using demographical and behavioral data to analyze which appeals will yield the most gifts in fundraising.
- Using internal and external economic indicators, to predict the price of shares in a company.
- Using previous consumption information to estimate the amount a customer will spend while visiting a store.
- Using historical trends to predict the demand for certain machines in manufacturing in the coming months.

In each of these cases, a model can describe the relationships between the data and the outcome. Models are built on a single table that contains all of the relevant information.

The Business Question

A business need should drive every model. The first step is to translate that need into a question that can be modeled. A good question has a specific set of data, a knowable outcome, and an actionable result in mind. Here are some example questions:

Mateo's Main Point: Predictive models should be driven by business needs, and you should know what actions you will take based on the results. Often people jump into building models way too quickly.

- Which customer segments are likely to respond to an advertising campaign?
- Which prospects are likely to make large donations?
- How effective is a new medicine at treating a certain disease?
- How many calls will a customer support center receive on a given day?

With each of these questions, there is a particular population or subset of data to be examined. There is also an outcome to predict that is present or can be represented within the data. This outcome has a clear potential actionable item as the answer. Without a knowable outcome to predict, we cannot train the model to identify the outcome. Finally, without a potential clear actionable result, the model becomes a "fun fact about the data" exercise. Rather than asking broad questions, it is better to take existing business needs and formulate them into actionable questions.

Linear Regression vs. Classification

There are two core modeling algorithms that can be used in most models: linear regression and classification. (The particular technique used for classification is "Logistic Regression".) There are a variety of other techniques that can be used for some specialized cases (such as neural networks), but these two are relatively easy to use and can solve most problems. These two types of regression are used for two different types of questions.

Mateo's Main Point: Linear predicts "How much?"

Classification predicts "Will?" or "How Likely?"

- **Linear regression** predicts a numerical value. Examples include predicting the temperatures in the coming week, pricing of various stocks and bonds, the expected unemployment rate in the coming months, or the agricultural output of a farm given the growing season.

- **Classification or logistic regression** looks at a binary outcome (success or failure, win or lose, will purchase a product or not, yes or no) in order to predict the likelihood of a future occurrence. Classification gets its name from its common use: classifying the population into two groups. One group is associated with a preferred outcome. In this case, we want to see how likely is someone to become a part of the preferred group in the future.

CASE STUDIES TABLE 1.1

LINEAR REGRESSION CASE STUDY: POOL SUPPLY CALL CENTER

A pool supply company has a customer service call center. The company would like to be able to forecast the number of calls on a given day. This will allow them to optimize the number of staff needed to answer the customers in a timely manner. The business question can be phrased as "For a given day, how many calls do we expect?" This has a specific set of data (the historical call center data), a knowable outcome (the number of calls), and a clear actionable result (adjusting the number of staff on the call center). This is a linear regression model because the outcomes (the number of calls) can take a range of numeric values.

CLASSIFICATION REGRESSION CASE STUDY: PROSPECT IDENTIFICATION

A university would like to find million-dollar donors and has many alumni in its database. The business question can be phrased as "Of the alumni, who are the most likely to donate a million dollars?" This is a classification model because the outcomes can be one of two things: either an individual will give a million dollars or more, or they will not.

NOTE: Technically this could also be run as a linear regression model, but trying to forecast a skewed distribution like this is hard—hence the preference to first forecast it as a "set membership" classification. The forecast will be skewed because the vast majority of alumni will never be able to give a million dollars, yet there will be a handful who will give amounts much larger than a million dollars. Distributions like this are hard to forecast, and we would recommend doing this as a second step, if at all, and comparing the results with the classification model.

Base Population

The base population consists of the people, items, or occurrences that are of interest in our model. The base population contains existing data used to provide insight into what is associated with our desired outcome. For a model predicting the temperature on given day, the base population is the historical data for previous days. For a model on whether a customer likely to respond to an advertising campaign, the base population would be all the customers who had previously been exposed to the campaign, some of whom responded and some who did not.

In the data, the base population is represented by the rows of the table the model is to be built on.

The base population should be made up of entities with similar experiences, which are capable of the same behavior. For example, if we have a list of donors but it includes individuals and corporations, the experiences and influences of these will be very different—that is, a corporation will not have a gender, cannot attend events, or respond to emails. So, we might want to build a model only on individual donors. Instead of looking at all the data in a table, we might look at a smaller subset of rows. Additionally, all entities in the base population should have the capability of exhibiting the outcome. For example, if we want to model voting in an election, we wouldn't want to include those under 18 in the base population.

The size of the base population must be large enough for the modeling algorithms to make statistically significant conclusions and extrapolate future predictions from past results. A safe rule of thumb is to have at least 30 rows per explanatory factor.

Mateo's Main Point: Use a subset of your data that look like the population whose behavior you want to predict.

CASE STUDIES TABLE 1.2

LINEAR REGRESSION CASE STUDY: POOL SUPPLY CALL CENTER

To build a model on the number of calls we expect in a day, we need a table with historical information at the day level, that is, a table which is constituted of one row per day, with columns for characteristics about that day. The base population is each of the historical days. In this scenario, we have 5 years of call center data. If we have multiple types of calls, we can build separate models for each type of call.

CLASSIFICATION REGRESSION CASE STUDY: PROSPECT IDENTIFICATION

To build a model on which alumni is most likely to donate, we will need data at the alumni level. Although the university has data on many affiliated entities, our base population will just be alumni. Non-alumni don't have class years or student activities and don't attend reunions. Thus, if we want to use those factors in our modeling, we will need to limit it to just alumni or else those factors will be biased against non-alumni since they are not able to have these experiences.

Target Field

The target field is a column in the data that describes the outcome to our question. For linear regression, the target field contains the values we would like to predict. For a classification model, you need a field with values of "0" or "1". You might have such a field in your data, or you may create one using an expression.

Mateo's Main Point: The target field represents the outcome to predict.

Doug's Helpful Tips: With some tools you can also build a classification model on the "selection state", the set of rows that have been graphically selected. You can conveniently do graphical selection to identify the interesting population, and then build a model to predict members of that population.

The target field is also known as the "dependent"

or "response" variable. It is the variable that is "dependent on" or "responds" to changes in the explanatory factors. For classification, it is convenient to use the term **Target Population** to describe the subset of data that exhibits the target behavior.

For a classification model, there must be an appropriate ratio between the target (value "1") and base population. The target-to-base ratio should be at least 1 to 200. If it is too small, then you will need to reduce the base population. For example, if forecasting major giving, then you can cut the base to people who are rated over $50K or something similar. You can still score everybody even if they are not in the base. In addition, there should be at least 30 rows relating to the favorable or positive outcome that we are trying to predict.

CASE STUDIES TABLE 1.3

LINEAR REGRESSION CASE STUDY: POOL SUPPLY CALL CENTER

For our call center, we will try to predict the number of calls received on each day. Thus, in our table, we need one column that contains the number of calls received in previous days.

CLASSIFICATION REGRESSION CASE STUDY: PROSPECT IDENTIFICATION

The target population should be alums who have given a million dollars or more to the university in the past. In the case of the university, there are 500 alumni who have given over a million dollars to the university out of a pool of 50,000 alumni. The ratio for this is within acceptable limits to ensure significant extrapolation from the target population.

Explanatory Factors

The explanatory factors, also called independent variables, are things that we consider to have a potential influence on the target outcome. Often, they come from hypotheses we have about the data and its relationship to the target. Each explanatory factor is a column in the table containing the base population. Explanatory factors can be a

Doug's Helpful Tips: A robust model will typically run with 10–15 explanatory factors. We sometimes see teams trying to work with far more—190 in one case. That's almost always a bad idea because the model will rarely have enough data for that and the result will suffer greatly. The technical term is "overfitting."

Mateo's Main Point: Explanatory factors are other data used to explain and predict the model target. Explanatory factors must be in the same table as the target field.

quantitative measurement or a categorical characteristic. Examples of quantitative explanatory factors include the income of a household, the amount of a treatment given, or an individual's height and weight. Examples of categorical factors include marital status, the type of advertising campaign (mail, email, telephone…), or the state of residence. Data for explanatory factors can also be brought in from multiple tables into the main modeling table.

One distinction to make is between "correlation" and "causation." The model results show correlations between the explanatory factors and the target field, and that may not always imply causation. "Correlation" means that two values vary together consistently. This correlation may be because the target is "caused" by the explanatory factor, but this is not necessarily true. See *Independent vs. dependent* and *Confounding and lurking factors* in the next section. We need to think through the explanatory factor and how they relate to the target in order to infer causation in the relationship.

Thinking Through Factors

It is important to think through and discuss the explanatory factors and identify potential sources of error and bias. Here are some additional considerations:

- **Names or keys**—Fields that uniquely identify an entity, such as a name or key IDs, are not useful for modeling. These are arbitrary, and you do not want to include them in models.
- **Independent vs. dependent**—Explanatory factors should not depend on or be caused by the target field. For example, if the target population is those who made a large donation, we do

not want to include attendance at a donor recognition event as a potential factor. Such a field would result in a model that perfectly predicts the target but does not tell us any information we do not know. In general, a perfect model is a bad model.

- **Binning**—Binning refers to grouping values in the explanatory factor. Binning applies to both quantitative and categorical factors. For quantitative factors, binning reduces the impact of noise, outliers, and highly skewed data. Quantitative factors can also be binned into groups that we reason a priori to have similar impacts on the target field. A common example of this is grouping age into certain demographic segments, that is, under 18, 18–35, and so on. If we make a chart on Customer Value vs. Age (see Figure 1.1), the data points would not create a straight line that increases from young to old. Binning the ages (see Figure 1.2) helps the model to capture the nonlinear impact. For categorical factors, binning assists in analyzing data with low counts in each category. For example, a doctor's specialty can be very specific and certain smaller specialties might have a low number of doctors. Binning the specialties into more general groups can ensure that there are enough observations in each group to have statistically significant results. Binning categorical factors also assists in reducing the number of possible values for large cardinality factors. Binning groups these miscellaneous

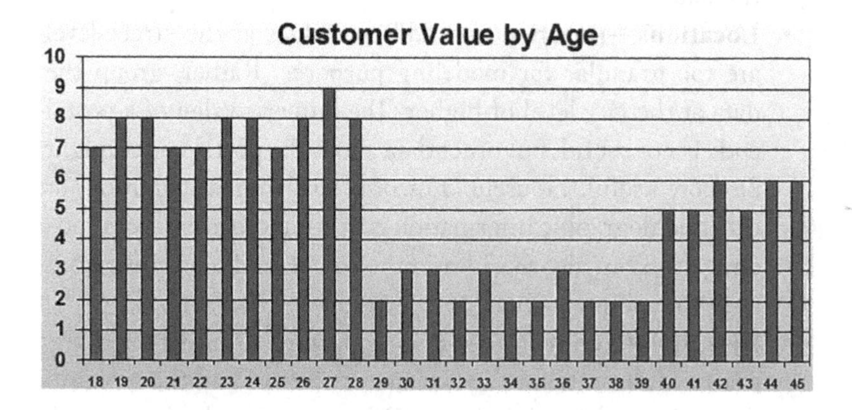

Figure 1.1 Customer value by age: data before binning.

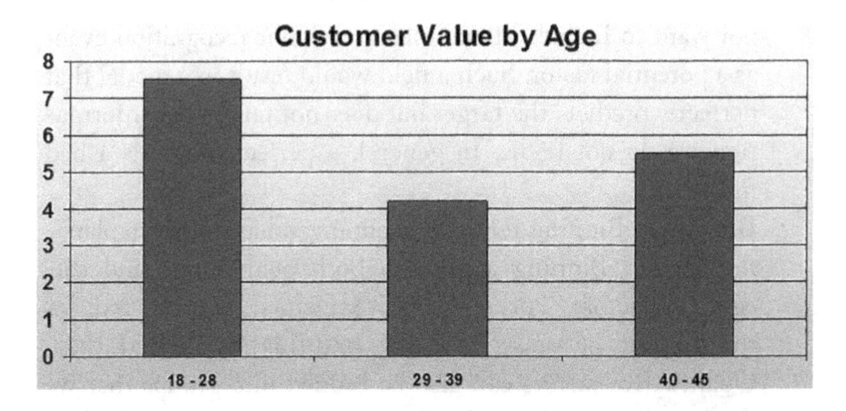

Figure 1.2 Customer value by age: data after binning.

bits of data into its own category. This increases the speed of modeling and might pick up on an otherwise unknown trend if each of the possible miscellaneous categories were considered on their own.

- **Dates**—Dates are not useful in a model since the relationship to a specific date in the past is not useful. It is better to look at certain characteristics of that date, such as day of the week or month. Consider whether there are cyclical trends by month or year. You can also compare two dates. Examples include the time since purchase, the time between multiple contacts, or the number of days from today. Many tools do not allow dates to be used at explanatory fields. Instead, you should create additional fields with the Date Parser based on the date.
- **Locations**—Location and address values at the street level are too granular for modeling purposes. Rather, group the data at the city level or higher. The numeric value of a postal code is not useful, but other data about the postal codes might be more useful. Household income level, age distribution, or other demographic information is much more useful for modeling. You can also use location to calculate the distance from a specific point, such as store location or major city.
- **Bias in data collection**—It is important to understand the source of the data for the explanatory factors and the biases in data collection it might bring. Often, we won't have complete data for a given explanatory factor; some of the rows might

have missing data. Most modeling tools can work with missing data. However, we should avoid a factor for which there is a systematic bias in collection. For example, we should avoid a factor that only a subset of the population has. As mentioned previously, it would be better to adjust the base population. Another example is survey data taken from an event will skew information toward those who attend events. Another source of bias is systematic measurement bias. This is bias that consistently offsets values by a certain amount. For example, a scale for weighing incoming shipments might always read off values by several pounds.

- **Confounding and lurking factors**—A confounding factor (Z) is one that is correlated with both the target field (Y) and another explanatory factor (X). The target field (Y) and the other explanatory factor (X) will appear to be correlated, when it is really the confounding factor (Z) that is correlated. A lurking factor is a factor that is not taken into account in the model. It has similar effects as a confounding factor. For example, if we look at incidents of drownings at the beach and ice cream sales at the beach from month to month, we would notice that the two are correlated. We might then infer that ice cream sales cause drownings. In reality there is a third factor, the season or temperature. More people are at the beach during summer, causing more ice cream sales and likelihood of drowning incidents. Not accounting for confounding or lurking factors can cause spurious conclusions about the data.
- **Multicollinearity**—Multicollinearity occurs when several factors are highly correlated with each other. This can cause overfitting, inaccurate assessments of the impact of each of the correlated factors, and high sensitivity to the correlated factors. The overall prediction will still be accurate, but weaker with regard to each individual factor. There are two options in dealing with multicollinearity. First, we can re-examine and remove one of the factors. After all, if two factors are highly correlated, we don't lose information taking one of them out. Alternatively, we can use a larger sample size or obtain more data. The factors might have just happened to be collinear for

that data set only, and larger data sets might reduce the correlation between them. Finally, we can choose to ignore the multicollinearity, keeping in mind the potential consequences above.

- **Outliers**—Outliers are data points that are far from the rest of the data and have extreme values. Outliers can come from bad data, unique one-off cases, or simply due to random chance. For example, a data entry error may cause an extra 0 to be typed in giving a value 10 times what is expected. There are several ways to adjust for outliers, although the matter is subjective. The first is to simply exclude the data point. If the reason for the outlier is clearly a data entry error, then excluding it makes sense. However, because some amount of outliers can be expected with large data sets, most would shy away from outright excluding. Alternatively, one can Winsor the data. This means taking the outlier value and replacing it with a less extreme value and trimming the range. For example, if most data points lie between 70 and 110, we would set any value beyond 110 to 110. Finally, one can keep the outlier. Some outliers are expected for large data sets, so removing them is unnecessary.
- **Overfitting**—Overfitting is when a model describes the data too specifically. Rather than describing large overall trends, the model describes every specific instance. Although it is extremely accurate for the existing data, it has poor predictive capabilities. Overfitting occurs when there are too many factors for the size of the target field.
- **External data**—Explanatory factors can come from other tables beside the main table used for modeling. You can bring data from external sources into the main modeling table. Often data will need to be aggregated in some way to get them into the same level of detail as the main modeling table. For example, data from a sales transaction table should be aggregated to the customer level and brought into the customers table. Sales data can be aggregated many ways, including counting the number of transactions per customer, the largest transaction per customer, or whether a customer has bought a certain product.

CASE STUDIES TABLE 1.4

LINEAR REGRESSION CASE STUDY: POOL SUPPLY CALL CENTER

To estimate the volume of calls in the call center on a given day, possible explanatory factors include factors related to business operations and factors related to the day. Factors related to other business operations include sales history, the introduction of a new product, or whether there is an advertising campaign underway. The day of the week or the month can have an impact. More calls might come in during the summer months as pool usage increases or on Mondays after problem is identified during the weekend.

CLASSIFICATION REGRESSION CASE STUDY: PROSPECT IDENTIFICATION

Possible explanatory factors for alumni likelihood to give include demographic information about the alumni, engagement in recent years, and information about their time at the university. Demographic factors include their age, their distance from the university, the median income of their ZIP code, or whether they are married to another alum. Behavioral activity could include attendance at reunions, the number of other events attended, or whether they buy season tickets to athletic events. Information about their time at the university could include their degree and major, what sports they played, and what student organizations they were a part of.

Model Training and Output

Once the modeling algorithms run against the training data set, the model is created. Training a model finds which explanatory factors correlate with the target field. It finds the magnitude and direction of the effect, whether positively or negatively correlated. These factors combine together into an equation that calculates the predicted target field value. For linear regression, this is the actual predicted value of the target field. For classification, the model calculates a score that corresponds to the likelihood of an individual belonging to the target population. This model equation can then be applied to new data to

Mateo's Main Point: The model shows which explanatory fields are correlated with the target field and how they are correlated, and it creates an equation that uses these to predict a target outcome.

Doug's Helpful Tips: At this point, context matters. Successful teams discuss together the model outputs, and they go through each influence factor and discuss whether the degree and direction of the influence makes sense.

And, this is always best done with the end-users who know the operation and the stories behind the data. It is critical to connect the math (model) to reality (the operations).

Unfortunately, often this does not happen.

predict future outcomes or to data outside of the initial base population.

Not all explanatory factors will be relevant to the model. Each explanatory factor is evaluated to determine whether its variation is statistically significant to the variation in the target field. There must be a large enough correlation between the change in the factor and change in the target. The statistical threshold for significance is known as the *p*-value. It is the percent chance that the result that we have in the data is due to random fluctuations and not due to an underlying trend. Common values include 0.1, 0.05, and 0.01, with smaller values indicating a smaller percent chance of random variation.

Each relevant explanatory factor is included in the model equation with a coefficient and possible transformation. The coefficient represents the change in the target field corresponding to one unit of change in the explanatory factor. Positive and negative coefficients indicate positive and negative correlations, respectively. Transformations account for nonlinear relationships such as factors that have diminishing marginal returns. Examples include taking the square root or the logarithm of a factor.

It is important to examine the model output and ensure that the results make sense. Go through each of the significant factors, and check the results against common sense. Try experimenting with other factors, iterate through the process, and continue to refine the model. Create some charts to display the fields, calculate different transformations of the factors, and discuss the results with the team.

CASE STUDIES TABLE 1.5

LINEAR REGRESSION CASE STUDY: POOL SUPPLY CALL CENTER

The model for the pool supply call center revealed that the call volume increases on Saturday through Monday, increases in the summer months, and with increased sales in the previous month.

CLASSIFICATION REGRESSION CASE STUDY: PROSPECT IDENTIFICATION

The model for the prospect identification found that alumni who have given in the last 5 years, who played sports during their time at the university, and who are in their 50s are the most likely to give to the new athletics center. Additionally, the probability increases with the number of events attended and the number of years they were season ticket holders.

Eight-Step Process

The eight-step process we recommend when building models is as follows:

1. Define the business question in terms of data
2. Select a target field and identify the subset of the data that is relevant to the questions ("Base" Population)
3. Visually explore and form hypotheses
4. Select explanatory fields
5. Train the model
6. Examine the model and iterate:
    a. Are the correlation coefficients (% concordance or R^2) in an appropriate range
    b. Discuss whether the explanatory fields that scored high make "sense". Are they
        i. Truly independent, or could they be another manifestation of the target (e.g., attending the President's Dinner scored high, but only people who donate a lot of money are invited)

ii. Causal or just coincidental (talk through WHY the factor should have influence)
iii. Driven by and perhaps a subset some other factor (e.g., county comes up high and so does city; then try one without the other and then reverse and see what happens)
iv. Etc.

c. If a factor scores really high, trying splitting into subcomponents:
   i. For example, "# Activity Participations in the Last 3 Years" scores really high …
   ii. … then run a second model against just the Activities Table and examine the range of Activities' influence—for example, some activities may be high (e.g., going to a reunion), some may be medium (e.g., going to a dinner), and some may actually be negative (e.g., interviewing high school kids, many of whom do not get in)
   iii. If this is the case, then maybe create "# High Activity Participations Last 3 Years", "# Neutral Activity Participations Last 3 Years", and "# Negative Activity Participations Last 3 Years"
   iv. And then rerun the main model with these three factors, vs. just one composite factor, and reassess the outcome
   v. If the original factor had a low overall score, then splitting like this would not be necessary and would have minimal impact on the overall model
d. Rerun the model multiple times and discuss again

7. Integrate the output into the core reporting/data discovery systems or tools for general use
8. Continually evaluate model performance

The model-building process should be a collaborative effort with discussions at each step among the team. Discuss the main business question the model will answer. Brainstorm and hypothesize possible explanatory factors. The team should consist of both individuals who know the data and those who will use the end result. The team should focus on answering questions and start simple.

It is better to start with the data that you have and then iterate and evolve the model. A simple model completed now with perhaps just

5, 10, or 15 explanatory factors is better than a complex model that takes forever and perhaps could even overfit the data. We have seen examples of teams trying to work with 190-factor fundraising models, when there really are only a dozen or so things that matter and not enough data to discern between many more factors than that—so the 15-factor model will actually outperform the 190-factor model and is MUCH easier to build and run.

Add new factors as they become available, and evaluate the model performance in the months after the model is built. Embed the model in your data discovery or reporting system so that it updates.

Clearly Present the Model to the Team

Remember that most nonstatistical team members will probably not relate well to modeling coefficients and numerical scores. So, we recommend binning the scores into five or so bins with very user friendly names. In fundraising, we might show Attachment Scores for a population in five bars, for example (see Figure 1.3).

This format enables end-users to have a discussion about their highly engaged prospects without getting into a debate about why one is a 0.72 and another a 0.64—which is often what happens if the raw scores are presented.

This being said, at some point in the discussion, it is relevant to show enough detail so the team can understand where the current score came from, and what could perhaps be done to change it. So, in the example above, we would also recommend having a page that has the details on each entity in the different groups, with the score and

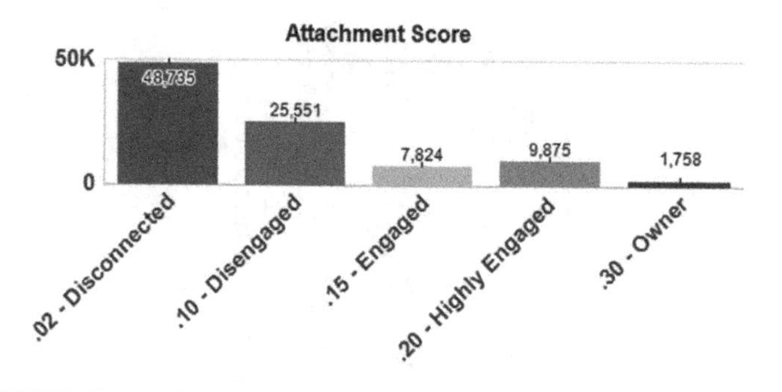

Figure 1.3 Attachment scores.

Attachment Score Details

ID	Name	Rating	Total Lifetime Commit ▼	Attachment Score ▼	Attachment Group	#CommitteesL10Y	#GiftsL5Y	#GiftsL6-10Y	#Reunions	#Sports	#Student Activities
26179	Matejko, Van	0 - NA	$3,495	0.20	.20 - Highly Engaged	1	5	5	0	0	4
188214	Viviers, Calandra	08 - $50K - $99K	$13,540	0.20	.20 - Highly Engaged	0	5	5	20	0	1
187903	Thongthai, Zachary	09 - $25K to $49K	$4,440	0.20	.20 - Highly Engaged	0	5	5	20	0	1
188962	Lawrie, Oscar	07 - $100K - $249K	$4,100	0.20	.20 - Highly Engaged	0	5	5	20	0	1
188300	Foster, Giuseppe	09 - $25K to $49K	$3,858	0.20	.20 - Highly Engaged	0	5	5	20	0	1
191888	Sorensen, Nigel	07 - $100K - $249K	$1,000	0.20	.20 - Highly Engaged	0	5	5	20	0	1
44720	Zhang, Deshaun	09 - $25K to $49K	$1,245	0.20	.20 - Highly Engaged	0	5	5	11	2	4
34596	Feuerstein, Aurelio	0 - NA	$16,900	0.20	.20 - Highly Engaged	0	5	5	13	1	3
29327	Sarma, Bernard	09 - $25K to $49K	$3,923	0.20	.20 - Highly Engaged	0	5	5	14	0	3
28706	Battistella, Graham	07 - $100K - $249K	$1,403	0.20	.20 - Highly Engaged	0	5	5	14	0	3
41429	Egan, Barney	08 - $50K - $99K	$16,510	0.20	.20 - Highly Engaged	0	5	5	12	1	3

Figure 1.4 Attachment score details.

also the factors that make up the score. For example, Figure 1.4 lists 11 highly engaged prospects.

This list provides the details on why the group is highly engaged in an easy-to-understand format. In this case, all have made donations in all of the last 5 years, and also the 5 years before that, and all have been to a lot of reunions and played some sports and activities. Why they are highly engaged is pretty clear. But what also stands out is one has been on a volunteer committee, and since the factors are arranged from left to right in order of priority, there is a clear opportunity to increase the Attachment Scores of this group by getting them on the right volunteer committee.

Conclusion

Building an effective predictive model should be as much a discussion as it is a model-building exercise. The most effective and most used models that we see are developed through an inclusive process that includes the following:

1. Discussing and agreeing on the key Business Questions to be answered
2. Choosing the Model Type that best addresses the business questions—typically either linear regression or classification
3. Discussing and defining the best Target, Base Population, and set of Explanatory Factors
4. Prepping the data for the Explanatory Factors
   a. Where are the data? Can you get more?
   b. Names or keys
   c. Independent or dependent
   d. Binning
   e. Dates
   f. Etc.
5. Training the model
6. Discussing the output with end-users; content the math to reality
7. Iterate the model, add/delete factors, re-run it, discuss the output again
8. Present the model to the user community in a friendly manner.

2

A Two-Day Course Outline for Teaching Analytics to Fundraising Professionals*

Lessons for Academia

MARIANNE M. PELLETIER

Staupell Analytics Group

Contents

* A Two-Day Course Outline for Teaching Analytics to Fundraising Professionals: Lessons for Academia, by Marianne M. Pelletier © 2019. Staupell, LLC. Printed with permission.

Introduction

When I used to watch "The New Yankee Workshop" with Norm Abrams (www.newyankee.com/), I was frustrated that Norm spent most of the program setting up the project—cutting pieces, sanding, and fitting the item together. I had to wait until the last few minutes of the show to see the final assembly, and only then would all the preparation steps made sense. I always wanted to start at the end of the program in order to make the beginning and the middle worthwhile.

Teaching analytics is like that: Analytics students can spend weeks learning data preparation, significance testing, checking for colinearity, and using the bell curve. And then, just as their learning muscles fatigue, they get to run a model to see the results.

The analytics courses I teach are often for operations staff who support fundraising operations. Analytics was introduced to nonprofits in the 1990s, and an entire profession has grown as a result. Now, an increasing number of fundraising practitioners are looking to expand their marketability and the services to their nonprofits by learning how to model prospects, mine their data, and use machine learning.

How does an analytics instructor, then, assure the success of these students, given that they work full time, have limited budgets, and work in a person-to-person industry? This chapter addresses one method—a two-day crash course. In this chapter, I will describe the context for the course, preparatory steps necessary, and a high-level description of the content, including a few examples of exercises, ending with takeaways that might be beneficial to those teaching these concepts in an academic environment.

Context

Fundraising analytics combines three separate disciplines, which must all be taught in tandem for the student to succeed:

- **Computer programming**: Like it or not, new variables have to be computed, and even the most sophisticated and easy-to-use statistics programs require some form of writing code.
- **Statistics, data mining, and machine learning**: These math and programming disciplines have their own languages and

methods, and, like learning to draw a piece of fruit before drawing a person, a student must study the bell curve before taking on methods like factorial analysis (description: https://en.wikipedia.org/wiki/Factor_analysis).
- **Fundraising**: The job of fundraising analytics is to answer a fundraising question, so a practitioner must understand the profession before launching an analytics project. If not, resources are lost examining the wrong data.

Since my students tend to come from the operations side of the fundraising business, a large portion of my students are more likely to be well versed in one of the three skills listed above. However, for those who are new to all three, a first class can be daunting.

Either way, beginners still have to work the modeling outline – exploratory statistics through to regression – in one class. If this full range of starter modeling techniques is not introduced in a single course, then students are not prepared to do modeling when they return to their offices.

So, how do we design a curriculum that meets these needs? There are three ways that I teach analytics to newcomers:

- Via a book or workbook
- Online, either prerecorded or via live webinar
- Classroom-style instruction or instruction in person

In this chapter, we explore answers to this question through one delivery method – the in-person class experience.

Before the Course Starts

Some preparatory steps are needed for any class. Here is the list for an intensive analytics course.

Understanding Your Audience

A successful course must accommodate all three learning styles, as defined by Fleming's (2001) Visual Auditory Kinesthetic (VAK) model (see here for a full article on the model: www.linkedin.com/pulse/online-quiz-3-different-ways-people-learn-vak-model-dannielle-walz/):

- **Seeing**: Students who are visually focused like to read first and then use graphs and colors to help their studies.
- **Hearing**: Students with an auditory focus learn best through listening to lectures and thinking with words rather than pictures.
- **Doing**: Students with a kinesthetic focus prefer to try out new concepts through exercises and learn through action.

Instructors have to teach to all three types of learners, and all at the same time. This variety of learning styles requires, then, a variety of materials and presentations:

- Reading materials, videos, graphs, and illustrations for the visual learners
- Verbal descriptions and allegories for the auditory learners
- Exercises for the kinesthetic learners

My own method is to send the handouts and supplemental materials to the class before it begins; review the concepts using both visual and auditory information during the class; and then take breaks during the class, assigning exercises to cement the concepts into the student's brain.

An Appropriate Classroom

Most professional workshops and conferences are held in hotel conference spaces with chairs lined up and nowhere to put down one's coffee or notebook. An analytics class best services its students when they can bring and plug in their laptops for the exercises and have space to work.

So, a good classroom is arranged with tables, plugs (I once had maintenance men scrambling through my class halfway through the first morning, supplying power strips), a white board or chalk board for drawing illustrations, and a projector and screen for the instructor's laptop. Because I also play videos, I need a sound system as well.

Well-Chosen Software

The course that I teach uses IBM's SPSS software because it is visually intuitive and looks like Microsoft Excel. Getting temporary licenses for this software, however, takes negotiation with IBM staff and then requires a process of sending the temporary licenses to each student as well as instructions on installation.

This extra step, however, is often supported by IBM sales staff since I am teaching the company's software to new users. However, courses can be taught in a variety of software, and my firm is developing courses using R.

Sample Data

Students are also sent instructions for downloading some of their own data for analysis, including which variables to include and hints on what kinds of records. Because employers require data security, students arrive with data on different media, and sometimes without data at all. For this last cohort, a sample data set on a flash drive is provided. The sample data set is also used as the demonstration data set, so students using it for exercises can see the expected results.

Examples of Real Situations

Students complain that generic examples had nothing to do with their professions. Who cares about the characteristics of different irises in the R sample data when the annual giving program is behind on its goal? In a fundraising analytics course, examples must come from the fundraising world.

I was once challenged to explain, "What Using Standard Deviation in Fundraising Can Do for You." At first I found the assignment preposterous. However, when I answered the question by showing that gift sizes in fundraising are so disparate that transformations are necessary, I saw the value in connecting every piece of course content to the student's experience.

A Workbook

The professionals who attend my courses are distracted. Their offices call with emergencies, and then they get bombarded with urgent work when they return. A workbook carries the class back with them so that, when they are ready to try a model on their own, they have a guide to help them pick up the thread again. A good workbook includes this content:

Written reviews of each concept covered in the course
Screenshots for each step, especially for more complex tasks (remember that the students get to try each task once or twice only during the course)
Examples of different uses for each technique
Exercises
A resource list for further reading
Open notes spaces

I am a big fan of the Dummies series of how-to books, so my workbooks also have additional technical tips and other inserts to catch the reader's eye. I also use three-ring binders for the workbooks and hand out extra three-hole punched paper so that students can tuck notes into the workbook pages throughout.

Hands-On Exercises

Trying out a new skill in a safe classroom is paramount to making sure that one can use it afterward. Discussion with PowerPoint slides alone often loses an audience, and getting lost at any step in the analytics process stops the learning cycle altogether.

Especially for kinesthetic learners, being able to try it out, whatever it is, makes sure that a learner gets to practice in a safe setting where mistakes have no consequences. For all learners, the Chinese proverb applies:

I hear and I forget.
I see and I remember.
I do and I understand.

Like learning a sport, learning analytics involves practicing techniques, making mistakes and fixing them, and then trying a new technique. For instance, exploratory statistics is often the first brand of statistics that people learn. Significance testing builds on exploratory statistics, and linear and logistic regression builds on significance testing. All are methods that must be practiced to be learned well. Exercises should be predesigned and include a question on practical application. See Figure 2.1 for an example.

Using·Mean·in·Fundraising¶

- Find·the·mean·lifetime·giving·for·all·donors·in·your·database·and·then·prospect·those·whose·lifetime·giving·is·above·that·value.¶
- Get·a·report·of·members·who·became·donors·in·the·last·full·fiscal·year·and·determine·their·mean·new·gift·(do·this·exercise·with·new·parents·or·new·donors·as·well).¶
- Find·the·mean·gift·to·your·direct·mail·campaign·when·you·do·not·make·a·specific·ask.¶
- Find·the·mean·major·gift·amount·per·capacity·rating.·For·instance,·if·you·have·100·prospects·rated·at·$100-$250K,·find·the·mean·major·gift·amount·of·those·who·gave.·Is·it·at·their·capacity?¶

Figure 2.1 Sample exercise from the exploratory statistics portion of the class.

An In-Person Modeling Course: Exploratory Statistics to Regression

To enable students to be fully able to conduct their own analytics projects back at their own office, a core group of skills must be taught during their first course. Some shorter courses offer only portions of this outline, but they should not be touted as complete analytics courses. Math classes usually start with algebra and end with calculus. In the same way, a new statistician/analytics professional cannot jump into factorial analysis before learning the basics of the bell curve: The results table would be entirely Greek to her. If someone is brand new to any type of analytics, then he or she should start with exploratory statistics, moving through significance testing and ending with linear and logistic regression. See the outline below for an example.

- Exploratory statistics
- Significance testing
- Linear and logistic regression
- Testing results
- Visualization
- Machine learning and data mining techniques can follow; but, like an art student learning how to draw a pear before drawing a person, an analytics student should always first learn the bell curve methods of modeling.

An advanced course on machine learning would still need to include a refresher on the bell curve before launching into Naïve Bayes methods or even decision tree algorithms. If students do not have the background to understand their results, they will abandon their new skills fairly quickly, so even exploratory statistics may have to be reviewed as part of even the most advanced classes. For instance, CHAID (chi-square automatic interaction detection) uses both degrees of freedom and p-values to do its work.

The following sections touch on each of the topics necessary for the intensive two-day course.

Exploratory Statistics

In the exploratory statistics section, we cover the best uses of measures of both central tendency and spread. For instance, Tables 2.1 and 2.2 show the effectiveness of looking deeper than totals when measuring campaign results. Table 2.1 shows the total giving by major for a sample college campaign, which is the usual way to look at this kind of data.

Table 2.2 shows what can be gleaned by digging deeper using exploratory statistics to expose which cohorts are giving more on average.

Table 2.1 Total Giving by Major Code for a Sample College Campaign

Arts and Sciences	$13,112.60
Curriculum Dev.	$14,262.50
Social Sciences	$23,769.67
Computing	$26,724.21
Marketing	$29,912.47
Science and Tech.	$30,835.50
English	$42,292.40
Accounting	$168,568.00
Bus. Administration	$209,055.22
Grand Total	$602,448.39

Table 2.2 Average Giving Per Donor by Major Code

English	$761
Finance	$863
Language Arts	$1,182
Curriculum Dev.	$1,196

Correlations

		Giving2015	Giving2014	Giving2013	Giving2012	Giving2011
Giving2015	Pearson Correlation	1	.567**	.808**	.888**	.586**
	Sig. (2-tailed)		.000	.000	.000	.000
	N	1506	1506	1506	1506	1506
Giving2014	Pearson Correlation	.567**	1	.861**	.799**	.678**
	Sig. (2-tailed)	.000		.000	.000	.000
	N	1506	1506	1506	1506	1506
Giving2013	Pearson Correlation	.808**	.861**	1	.926**	.842**
	Sig. (2-tailed)	.000	.000		.000	.000
	N	1506	1506	1506	1506	1506
Giving2012	Pearson Correlation	.888**	.799**	.926**	1	.814**
	Sig. (2-tailed)	.000	.000	.000		.000
	N	1506	1506	1506	1506	1506
Giving2011	Pearson Correlation	.586**	.678**	.842**	.814**	1
	Sig. (2-tailed)	.000	.000	.000	.000	
	N	1506	1506	1506	1506	1506

**. Correlation is significant at the 0.01 level (2-tailed).

Figure 2.2 IBM SPSS output for correlation among giving levels in 2015, 2014, 2013, 2012, and 2011.

Significance Testing

Because fundraising focuses on raising gifts, gift sizes and gift counts are most frequently used in donor modeling and in nonprofit data mining. Given that, we teach significance testing not only to reduce the number of variables that will be tested in the linear and logistic regression models (in our consulting work, we sometimes test nearly 1,000 variables and computed variables) but also to prevent colinearity. The correlations test in Figure 2.2 shows how we illustrate using significance testing to find the best variables for the modeling from among colinear variables.

Notice that giving in 2014, 2013, and 2012 are highly correlated to the dependent variable, giving in 2015; however, these variables are also highly correlated to each other. Through this significance testing exercise, students learn to identify colinearity and then to identify the right independent variable for the final model. With this example, we often coach students to use the most recent variable—in this case, giving in 2014.

Linear and Logistic Regression

Once students have understood correlation and avoiding colinearity, their results in linear regression must also be understood.

Classification Table[a,b]

			Predicted		
			DonorFlag		Percentage Correct
	Observed		0	1	
Step 0	DonorFlag	0	0	432	.0
		1	0	1074	100.0
	Overall Percentage				71.3

a. Constant is included in the model.

b. The cut value is .500

Figure 2.3 IBM SPSS classification table serving as a confusion matrix.

Discerning how *p*-values, the *F* statistic, and R^2 all work together to frame the value of a model is the most important step to using the model well. We often discuss "good enough" around p-values and the Fs statistic and then have to shift to "better than before" around R Squared.

Then students have to shift in logistic regression to understand the confusion matrix. Although most software programs offer some form of a pseudo R Squared value, students are walked through using the confusion matrix. Figure 2.3 shows how a confusion matrix that is 71.3% correct overall can still represent a poor model, since it categorizes every record as donors.

Testing Results

Students then learn both how to compute the resulting scores in their databases by computing them in SPSS and then how to run a test table to see how the predicted scores compare. The key is not only to discard those scores that turn out to be wildly different from actual results (in this case, giving) but also to identify which level of probability from the logistic regression works best for identifying likely prospects.

Figure 2.4 shows that a predicted probability of 0.60 or better is the best cutoff point for identifying the data's best donors, since all donor records carry a prediction of 0.60 or more.

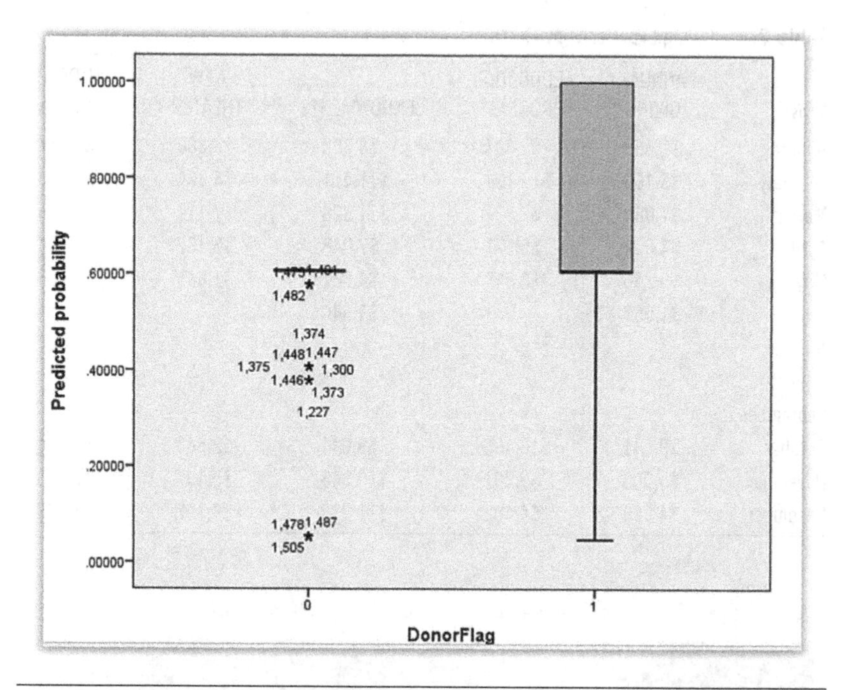

Figure 2.4 Boxplot from SPSS showing the floor of the predicted probability for actual donors, which sets the minimum threshold for choosing prospects who are most likely to give.

Visualization

Statistics and machine learning, surprisingly, are visual sciences. Data lined up, aggregated, or charted show patterns instead of the sums or averages expected in other parts of mathematics. For instance, in Table 2.3, one can clearly see the white space in August, which begs the question, "Why aren't prospects giving in the summer?"

Especially during the exploratory statistics and the visualization portions of the course, graphs bring the point home. For example, the tree map in Figure 2.5 is all the rage now that pie charts are out of style. It may look interesting and informative to a practiced statistician but may look like just an Impressionist subway painting to a first-time viewer.

My practiced eye notices that the trustees make the highest percentage of pledges to asks when asked to support endowments, giving about 31% of what they are asked to give. Corporations, however, respond with less enthusiasm to education programs.

Table 2.3 Giving Experience by Month

MONTH	ANNUAL GIVING	EDUCATION PROGRAMS	ENDOWMENTS	NEW THEATER	OUTREACH PROGRAMS
January	$7,669	$8,407	$8,308	$3,286	$6,917
February	$5,466	$10,114	$11,451	$4,304	$4,813
March	$7,080	$7,327	$11,228	$7,115	$5,845
April	$4,592	$5,571	$7,048	$6,102	$5,584
May	$5,260	$2,843	$4,837	$1,643	$2,425
June	$1,057		$1,540		
July					
August					
September					
October	$9,941	$16,466	$9,044	$2,645	$1,964
November	$5,005	$9,201	$10,348	$5,497	$7,640
December	$4,986	$4,402	$11,232	$8,192	$4,208

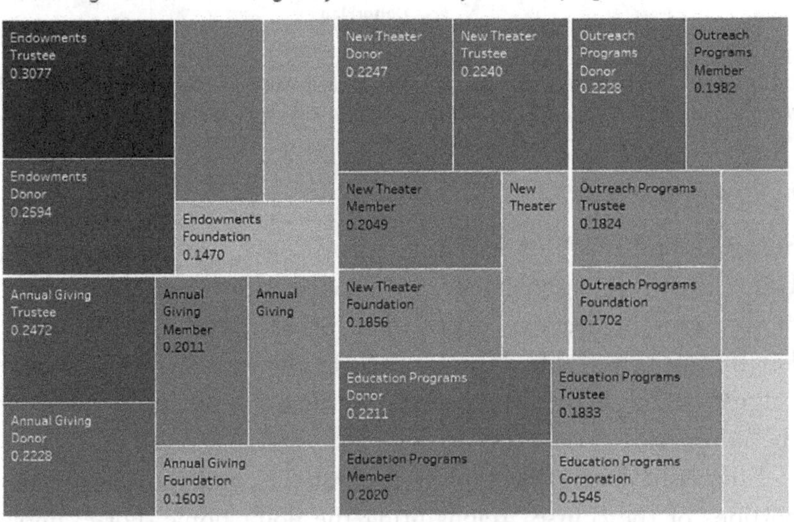

Figure 2.5 Tree map of percentage of asks vs. pledges by constituency and campaign.

What a fundraiser (the audience for visualizations like this) would more likely see is a bunch of boxes, some not labeled. Actionable insights are not evident to an audience not oriented to the data discipline. Figure 2.6, passé as it would be to us data scientists and other Tufte fans, is an illustration that would get the point across more easily.

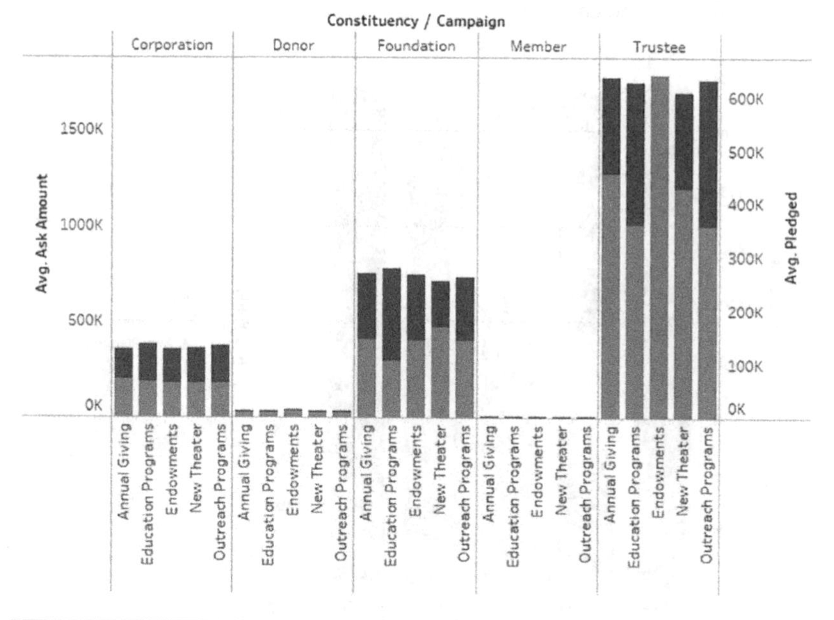

Figure 2.6 Bar chart of percentage of asks vs. pledges by constituency and campaign.

Though busy, this chart shows the relative success of asks vs. pledges during trustee solicitations for endowment gifts. Note also that the 31% from the tree map is computed record by record and that this chart is computed in aggregate, making the illustration entirely different.

A cleaner chart, though, would look at each constituency separately, as illustrated in Figure 2.7, which is filtered by Constituency (in this case, showing only data for trustees).

Conclusion

As an undergraduate, and even as a graduate student, my intensity taking focusing on the classes was much lower than in professional workshops. My empathy for professionals who are in the same boat pushes me to hone my course outline down to a specific list of skills that can be learned and practiced in the short time that I have my audience. When they go home, though, they have to know enough to conduct analytics on their own. Striking that balance is a constant effort.

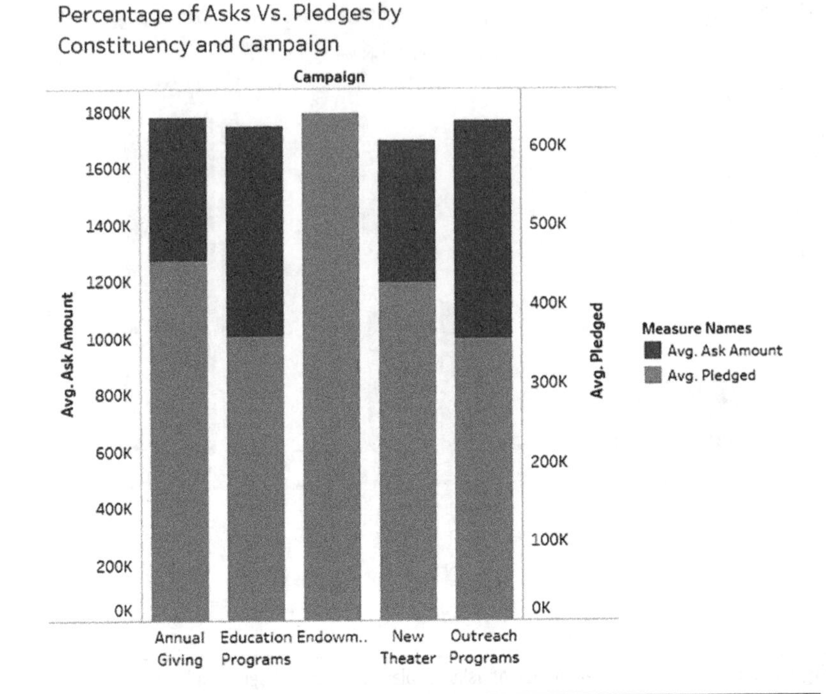

Figure 2.7 Filtered bar chart of percentage of asks vs. pledges by constituency and campaign.

Lessons learned from teaching professionals can also be applied to teaching analytics in the college or university classroom:

- Keep in mind that students may have different backgrounds, some coming from technical majors and some from functional majors. It's important to fill in the gaps for all of them.
- Even though your students will want to get to "the good stuff" immediately, don't gloss over key foundational statistics concepts such as the bell curve.
- Don't let the content drive the way the course is taught. Be sure to include material that reaches out to all three types of learners.
- Ensure that the physical environment is suitable for student learning.
- As much as possible, use data sets that are relevant for a wide variety of student backgrounds.
- Use workbooks or similar devices to make it easy for students to take good notes that will allow them to replicate in-class demonstrations.

Finally, a critical lesson that students need to take away with them is to always be aware of how familiar the audience is with analytics methods. While sophisticated tools are useful in discovering knowledge hidden in data, these tools may not be the best vehicles for supporting the narrative when briefing an audience that does not possess higher-level analytics skills. In other words, don't let the tools get in the way of conveying the message.

3

Developing Professional Skills in a Data Analytics Classroom

KATHRYN S. BERKOW

University of Delaware

Contents

Excellence in teaching data analytics is a moving goal post, as teaching any practicing discipline will always be. There are so many topics, teaching styles, and programming languages we can choose to incorporate. If we try to include too much, we could overwhelm and alienate students from this exciting, developing field. But if we focus too much on just one method, we risk underselling its evolving nature. Many would say it is also our responsibility as instructors to prepare students for their next steps by exposing them early to the less academic—but extremely valuable—skills they'll need after graduation. Of course this is true in many courses, but industry problems, methods, and technology are ever-present in data analytics, so it is natural to weave them into our curriculum.

Following graduate school, I worked in financial services and enjoyed the fast-paced, iterative problem-solving that accompanies product design and sales. On the job, I extended my understanding of data collection, modeling, and the importance of knowing your data and context. But the soft skills that came from working in industry may have the greatest potential benefit to my students—communication skills, flexibility, and knowing when and how to best contribute to a particular project or team. Each semester, I try to bring these fundamental characteristics of a great job candidate into the classroom and explain how the skills they're learning translate to resume items and job interview questions. The focus of my role at the University of Delaware is bringing an industry-focused analytical framework to undergraduate business majors with energy, enthusiasm, and business applications. My industry experience shapes the way I select and deliver material for my students every day.

Business analytics courses like mine often cover statistical testing for continuous variables, proportions, categorical variables, and the basics of regression and other models. But a surprising number of statistical methods can be rolled into linear regression—understanding relationships, correlation, comparing groups, within- and between-group variation, and of course the benefits of constructing a model for understanding behaviors of a system and proposing business solutions. So, my course covers statistical models exclusively, incorporating a continuum of techniques via linear regression. I add logistic regression, decision trees, and comparisons of models to provide a few more basic tools. I also introduce visualization, performance assessment, cross-validation, assumptions, overfitting, and other issues that may arise for analysts. I hope that when they leave the classroom, my students know that data is powerful—and that it's easy to find the tools they need to customize methods to new data sets.

Most students know that "big data" is an important phrase they should know when they hit the job market. But many don't know what that means—simultaneously everything and nothing—or how to discuss it with confidence. And while analytical skills are valuable, they're not absolutely required to become a productive employee after graduation. The effectiveness of an employee has so much to do with communication skills, an ability to work in an environment of constant change, and appropriate contribution—strong teamwork skills, independent learning, intellectual curiosity, ownership, individual accountability, and giving and receiving constructive feedback. I try to facilitate the development of these skills along with "big data" skills as much as possible. The remainder of this chapter covers these underlying course components and some ideas for how to work them into analytics courses. I also include a brief introduction to Team-Based Learning (TBL) [1] and how it may enhance some of these skills very naturally. The themes and methods discussed here are focused toward undergraduate students, but could certainly be tailored to fit a graduate audience.

Elements of Professionalism

While every introductory data analytics class focuses on data, methods, and meeting the two with some light programming, a practicing discipline like this one can benefit students by introducing a few competencies they will need for future roles in industry—analytical or otherwise. For me, the key areas I want to cover include communication skills, openness to change, and appropriate contribution to a team or project.

My students are generally sophomores with a variety of business majors and varying levels of interest in analytics and programming. On the first day of class, I underscore my awareness that not everyone wants to be a data scientist—it takes all kinds to run an organization. I give examples from my experience of how a variety of people interact with data scientists and how useful analytics can be in different types of roles. My examples range from sales to human resources, but I also point out that roles of the future haven't yet been imagined. By learning to appreciate analytics and the types of questions that can be explored, I hope they'll be ready to plug into a variety of entry-level roles and take exciting opportunities as they come along. That first discussion sets the stage for both the technical skills and underlying themes of the course—communication about analytics, openness to change, and managing our contributions—that make us all better employees.

Communication

Communication turns statistical calculations into business solutions. Anyone can learn how to calculate a value or create a linear regression model, but fewer can effectively communicate the meaning and value to an audience who can act on it. Communication in a data analytics class can be broken into two parts—converting calculations we perform on datasets into meaningful solutions and conveying them to audiences that can act on those results. Some people are more natural communicators than others, but we all need to work on these skills. Being able to communicate motivation, methods, issues, and results helps in development of business solutions, so each is an important skill for students to cultivate.

I often illustrate communication skills with the example of an interview, asking a candidate to explain a favorite analytical project. While it seems like a simple question, the way the interviewee answers it immediately illustrates their communication ability. If the answer contains the motivating research question, it's a very positive sign. Quick movement through the entire analysis from goals to conclusions indicates that this person will be able to hit the ground running. Students are dissatisfied to hear that the part I am most interested in is the "easy" part and not the technical part, but it highlights how much communication skills matter to employers. This is just one example of how communication will matter in their future, but we have all had experiences that illustrate this—a poor presentation or discussion that didn't meet its goals for some reason—and realistic examples may help motivate students to avoid pitfalls caused by poor communication.

Once students can think through analytical processes, they can begin discussing their results. This may happen with a technical audience like a manager or peer, or perhaps with clients or executives ranging the continuum of technical skills and interests. So, we also talk about audience and understanding how statistically motivated and/or results-focused a conversation should be (e.g. statistical significance vs. economic significance). I also give examples of "leading with the headline" in case we only get a few minutes to convince someone of the value of our results. And of course, preparing to communicate with a variety of audiences requires that students have confidence and ownership of the technical skills, too.

Because communication is such an important component of the course, I have experimented with different types of practice and assessment—though I'm sure there are many other effective techniques.

The two methods I've had the most success with so far are open-ended exam questions and a semester-end team project with an informal presentation component.

In open-ended, short answer exam questions, students must communicate technical results and summarize their impact briefly and completely. I might ask students whether they would recommend using this model going forward based on its performance, for example. Here, they'll make judgments about a model's stability and accuracy based on the business context and provide evidence for their decisions. This kind of question is "easy" because it does not require calculations but can convey exactly how well a student understands the material.

One of the questions I get most often is how answers will be graded when they are highly subjective. This is a great question, and my response is that this is the nature of their future roles. CEOs of two different companies could look at the same data and make different conclusions based on their unique perspectives. While students often remain frustrated without official "correct" answers, this conversation supports another idea within analytics—that we are searching for the truth in data, for which there is no "right answer" until after we have seen the results play out over time.

In addition to exam questions, I have also used team projects to evaluate students' analytical and communication skills. Students form groups and analyze a data set—investigate relationships in the data, build a model, consider outliers, assumptions, overfitting, performance, and make suggestions for next steps. Finally, teams visit my office for a casual, but graded, conversation about their motivation, methods, and results.

While there is great value in formal presentation practice with slides, I want the students to practice communicating analytical results to someone who is interested and invested, but also aggressive about asking questions just like their managers in the near future. Teams do struggle with typical issues that come from working together outside of class, but overall, they practice their skills and add a full-scale analysis to their resumes. Individual appointments with teams may not be possible in every class, but other options like informal presentations to the class or virtual presentations (either synchronous or asynchronous) could be effective alternatives.

Open-ended questions and informal presentations are just two methods of building communication skills that I have tried within a lecture-based style of teaching. Both have helped in identifying where misunderstandings of the material may be and how I can address them to improve understanding. However, as discussed later in this chapter, employing TBL in the classroom has amplified the role of communication throughout the semester.

Change

Aside from communication skills, I also try to help prepare students for the constantly changing environment they will face in industry. So many things change in industry roles every day, from the people around us to the needs of customers or the political and economic environments firms face. Managers and peers change, expectations and goals change, products and services change, and so do the analytical methods and technology we use. Of course, change can be incredibly stressful, but I remind students that changes often create opportunities to take on new responsibilities. Well-rounded employees must be ready and willing to change, as the first employees to arrive at the new normal are often the most successful. The sooner students become comfortable with change, the better, and with practice we can all improve our flexibility.

As in other areas, I share stories of change from analytical investigations during my time in industry with students. I talk about being surprised by a change during an analysis, learning of an important detail late in the game, losing or gaining a key stakeholder, or getting a new manager with different priorities or methods. I also mention learning new technologies to accommodate each new role. Some of my colleagues who learned to program in Fortran or Perl use this as an anecdote in class to illustrate how technology is continually evolving. Students are often surprised by these examples and fear that what they are learning now will be outdated soon, but I point out that the underlying skills and questions often stay the same and we can adapt thoughtfully over time. R and the basic model types we learn in class may not always be in fashion, but the next language or model they learn will come more easily and more speedily. And the questions they will learn to ask about model quality and results can be applied to almost any future analysis.

For change on a smaller scale, we learn extensions of basic models as the semester progresses. When we are comfortable with one technique we move on to the next, compare performance, and assess strengths and weaknesses. I don't want students to leave the classroom thinking they have learned all they will need in linear regression and the programming tool R, but rather to have confidence that they can learn to build any model or program in any language now that they have seen what is possible.

Finally, I also incorporate additional technologies in class for further exposure to change. In past semesters, we have learned how to capture data from Bloomberg for a basic example of data collection. In future iterations of the class, I'm hoping to introduce additional computing tools like SAS or Python. I'd like for students to see that we can get the same results from more than one tool, since tools and processes vary from firm to firm and even team to team. Of course, acclimating students to change is not something an instructor can do in one semester, but sharing stories of our own experiences and providing examples of small-scale change in the classroom provides at least some exposure. Later, I'll also share the ways TBL has contributed to students' flexibility.

Contribution

The final group of industry skills I attempt to overlay in class focus on contributions to a project or team. The skills in this group are unrelated to statistical modeling, analytics, or communicating results but are highly correlated with being a good employee and becoming a strong leader. This group includes independent learning skills, excellence in work as an individual or part of a team, giving and receiving constructive feedback, and accountability for one's own performance. These skills are useful in any role—in industry, academia, as a graduate student, employee, or entrepreneur—and can be developed if given some attention.

These important professional skills are tricky to develop in a traditional lecture-style class where students are less active, however. As usual, students' individual understanding can be measured via homework and exams and independent learning skills might be implicitly developed by students at home using their lecture notes to apply to homework tasks or prepare for exams. Teamwork skills might be practiced during the out-of-class team projects discussed above, but the results of team projects often appear to be disappointing to the students for a variety of reasons. They may have trouble arranging meetings and usually find a few teammates less helpful than others. Often at presentation time, there is a palpable sense of dissatisfaction with the results and the experience.

When it comes to feedback, students are accustomed to receiving interim grades in a lecture-based course and may use them to improve their performance in future assignments. However, giving feedback is something students do not often get exposure to early in their studies. In past semesters, I have experimented with adding a peer evaluation component to team projects; students evaluate peers' contributions so I can more fairly assign project grades and they gain exposure to providing constructive feedback.

Another way to illustrate professional feedback is to ask students to comment on the style and content of the course. Each semester, I ask students to evaluate the class informally in addition to the formal course evaluations at the end. I ask for suggestions in an online survey after the first exam, which helps me to understand students' experiences and learn what things could be improved, big and small. The students often have great suggestions that I can incorporate immediately or in future semesters. I present common themes in the responses and outline changes I plan to make in the short- and long-term that might help improve their experience in class. I hope that through mid- and end-of-semester course evaluations, students see that giving and receiving feedback is common, professional, and can benefit everyone involved.

As in many courses, students are held accountable in this class via attendance, participation, assignments, projects, and exams. As employees, they will be accountable for outstanding work, often reflected in performance reviews, salaries, raises, bonuses, and promotions. But accountability is sometimes lost in translation because students are accountable to many different grading schemas in their various courses simultaneously. In contrast, when we work for one manager—even if we have responsibilities to others—that person evaluates all of our work, helps balance our workload, and adds or removes expectations as needed. This might be an area where students have it easier after they graduate, because having just one manager makes it easier to meet and exceed expectations. Regardless, attendance and performance remain typical methods for measuring and rewarding student accountability.

Clearly, important skills like independent learning, individual and team accountability and performance, and giving constructive feedback get little exposure during a lecture-based version of an analytics course. In order to make the course more engaging and incorporate more opportunities to exercise all of their professional skills, I have recently started experimenting with TBL in the classroom.

Team-Based Learning

TBL was first developed in the 1970s by Larry Michaelsen as he was preparing materials for his undergraduate course [1]. He was frustrated with challenges around getting large sections of undergraduate students to come to class prepared, participate, and learn interactively. With some experimentation, he developed a new format intended to fix some of the typical issues with having students engage in teamwork while providing a more interactive experience in the classroom. My intention in the following section is only to introduce the motivation and methods of TBL and point interested parties to additional resources like Sibley and Ostafichuk's *Getting Started with Team-Based Learning* [1]. My goal has been to use TBL as a tool for incorporating communication, change, and professional skills into students' data analytics training.

There are a few important components of TBL that make it different from other classroom styles. TBL is a classroom paradigm with a focus on the instructor as facilitator, mentor, or coach rather than lecturer, guiding students through learning the material rather than delivering it directly. There are two different types of tasks in the TBL paradigm—team tasks and individual student tasks. Students may be assessed individually on items like homework, quizzes, and exams, but are assessed as teams on special readiness quizzes and on team exercises or projects. TBL literature suggests that students be assigned to teams of 5–7 students on the first day of class and keep the same teams throughout the semester. All (or most) of the work students do in their teams happens in the classroom, rather than at home, which generally reduces the overall stress and friction of working in teams. TBL literature suggests that care be taken in team selection to optimize productivity and learning [2].

Sibley and Ostafichuk recommend organizing TBL classes into a series of 4–8 modules [1]—groups of topics best digested together. Prior to the start of each module, students are tasked with learning some amount of material on their own at home. Each module begins with a Readiness Assessment Process that includes an individual quiz, the same quiz taken as a team, an appeals process, and a review of any questions missed by a large number of students. The students are not being asked to learn all of the material at home, only the major themes they will need to begin the work of the module. The Readiness process is recommended to take about one full class to complete, depending on the length of each class session.

When the Readiness process is over the material in the module begins, consisting of Application Exercises governed by the 4S's. The 4S's are Significant problem, Same problem, Specific choice, and Simultaneous report, the TBL principles of designing classroom exercises to guide students through the learning process. The idea is that each exercise should require each team to focus on the same, meaningful problem, for which they are asked to choose an outcome and all teams report simultaneously. Style of outcomes can vary and reporting methods can vary accordingly; Sibley and Ostafichuk provide lots of helpful suggestions [1]. The most general example would be giving students a particular situation, asking each team to choose the single best answer to a question about handling that situation, and then have all teams reveal their answer (e.g., A, B, or C) at the same time so all can see the agreement or disagreement of answers. If all teams agree, a discussion could follow where teams give detail about how they arrived at this decision. If all teams do not agree, then naturally a more exciting discussion can follow where students can hear the findings of others that led to alternative choices. Detailed examples and suggestions for how to design effective classroom exercises can be found in Sibley and Ostafichuk's text.

After working through a number of increasingly complex and decreasingly structured application exercises to cover the material in a module, the module can end with some type of assessment—team or individual projects, papers, or exams, for example. After the assessment, the process begins again with new at-home study material and the Readiness Assessment Process.

The last of the major components of a TBL class is a focus on peer assessment. The Readiness Assessment Process holds individual students accountable for preparation for a module, but the peer assessment component holds individuals accountable to their teams. There is a deep discussion in the Sibley and Ostafichuk's text on options for incorporating peer assessments into student grades and how to manage those assessments, but the idea is that if students are working with the same team all semester, we should provide a safe, high-quality environment to develop their teamwork skills with formative assessments and have accountability with some portion of the final grade decided by peers [1].

It is the combination of the Readiness Assessment Process, 4S activities, end-of-module assessments, and peer assessment that make up the TBL paradigm. Research aimed at determining the effectiveness of interactive classroom styles does seem to show that students absorb more information, attend more often, and show improved performance on assessments [3,4], even in large classes [5]. As TBL supports a more interactive classroom experience, instructors who try it may see some of these improvements in their own classes. Sibley and Ostafichuk's text presents a great deal more theory and evidentiary support around why TBL may improve performance in these areas [1].

I have incorporated some of the components of TBL into my data analytics classes in order to improve the overall student experience, and I have a few takeaways to share on its helpfulness in fostering industry readiness.

How TBL Facilitates Professionalism

Implementing components of TBL into the data analytics classroom has the potential to support development of important nontechnical skills as students prepare for future roles in industry.

First, TBL can help students practice their communication skills in a variety of ways. Because students are working with teams, they get more practice interacting with others working toward the same goal. This is a great way to practice for on-the-job project meetings and presentations. In a TBL setting, students read or listen to a prompt and come together to discuss an action plan. Because reporting of results happens very quickly, students are motivated to speak up with ideas for how to make a decision about the best solution or steps to find one. It is a realistic representation of a working environment, as opposed to an out-of-class project where some students may hang back to wait for instructions (or answers) from others. New hires quickly find that even teamwork is competitive in an industry setting, and that contributions are expected from everyone. Navigating contributions from team members and deciding which ideas to investigate can help them prepare for the same types of conversations with colleagues in their future—more realistic than in a typical group project. As a result of TBL, I've been able to replace our usual end-of-semester team project with regular group interactions and in-class, graded group assignments that generate opportunities to practice teamwork while eliminating some of the most frustrating aspects of teamwork for students.

Aside from communication with team members, a TBL exercise also gets students to practice defending their ideas live in the classroom in front of their peers and their instructor. After teams report their decisions, individual students are expected to speak in support of their team's decision in a group discussion. While intimidating at first, engagement with the classroom builds confidence speaking in front of an audience and requires students to plan what they might say and think through development of their evidence as they're working. This casual but high-stakes reporting environment is extremely realistic. For new employees, every interaction with a manager or stakeholder will be an informal opportunity to show value and build trust. TBL discussions may be perfectly sized opportunities to practice speaking confidently in front of a room genuinely interested in hearing the rationale behind an answer. And practice generates improvement.

During the team discussions and through defense of teams' decisions, students must also shift conversation styles from technical support of their methodology to interpretation of the impact of that business solution. This is great practice at quickly transitioning from analytics to impact and back during a conversation with a stakeholder at work. Students also get practice in this setting communicating with different levels of understanding in their audience, from those who completely understand their argument to those who are being lifted to their level of understanding *during* their discussion *because* of their discussion. In the best of these discussions, I see the additional TBL benefit of students teaching their classmates about the topics being covered.

In an effort to warm students up to welcoming changes in methods, technologies, roles, and goals, modified classroom interaction is a big change in itself. Students have come to expect traditional lecture-style course formats and a shift to TBL is a shift toward more accountability and active participation during class. Hearing this on the first day of class is enough to get some students excited and put others on edge about new expectations. While this is to be expected, it is also good practice in adjusting to change. Modeling challenges in the classroom that students will face on the job may prepare them to face those challenges with positivity. TBL is an exercise in handling change since it edits the expectations, interactions with peers, ownership and engagement, homework, and depth of understanding students may have in a course. It also requires constant transitions between lecture, discussion, and team exercises. Blending TBL into my class has added exciting variety and encouraged students to adapt.

Finally, TBL has created new opportunities for students to practice learning independently, giving and receiving feedback, and developing a sense of accountability for learning the material and contributing to a team. Every employer is looking for new hires to be motivated to learn quickly and independently on the job and grow the skills they have into the skills they need. TBL expects students to digest some material independently and then come to class for practice and discussion. As a result, students get exposure to learning from helpful resources. So many new hires report that they learned more in their first six months at work than they did in college. But as instructors, we know that college is about learning *how* to learn. By the time new hires shift from students to employees, they're trained to ask questions and learn methods and context in new settings. TBL encourages students to learn the basics of new topics at home by reading, listening, and watching instructor-selected resources, and this is just another benefit of the style. Practicing learning from resources like textbooks and websites is so valuable for students' futures—whether industry, nonprofit work, a startup, or graduate school lies ahead.

Because TBL offers somewhat realistic team experiences, students also get an opportunity to practice giving and receiving targeted feedback to and from key stakeholders. In the TBL classroom, team members and instructor often provide formal and informal feedback to help drive success. In future roles, students will be giving and receiving feedback from peers and managers. Of course we hope this will be a positive experience, but there are always points for development. When facing important feedback, it may help new employees to have been through this process before in a controlled environment as students. TBL peer feedback comes via peer assessments that can be formative, summative, or both, and can be incorporated to impact student grades in a variety of ways [1]. We all need practice receiving feedback gracefully, identifying things we do well, and those we can improve. This is a feature that is somewhat unique to the TBL paradigm because TBL keeps the same students teamed up all semester, making teammates stakeholders in each individual's successes or failures.

While traditional mechanisms like homework and exams still help instructors to understand individual students' contributions in TBL, student accountability can also be strengthened in a TBL setting. The use of TBL adds accountability via the Readiness process and ongoing teamwork. TBL encourages student preparedness by asking students to come to class ready for each module's Readiness quiz. Students not only perform on quizzes individually but as part of their team, so their peer group is also holding them accountable for preparedness. During team exercises, students who have read or watched the material, listened during explanations, and digested previous exercises will stand out to their peers as truly prepared, excellent performers. And expectation of receiving the peer feedback described above keeps students focused on contributing during team exercises. This type of accountability is exactly the kind students will see when they transition to employment—consistent accountability to peers, managers, and customers—so TBL could be a good way to enhance those skills before starting in their first roles.

Outcomes and Conclusions

As an industry professional, I found communication skills, adaptability, and the ability to contribute appropriately to be extremely valuable when combined with technical skill. Infusing these themes into the classroom will better prepare students for their next roles as new employees. Currently, a blend of lectures and TBL is working well for my course. I use short lectures to introduce concepts, and then TBL exercises to practice and extend skills during class. I use online polls to collect teams' answers to exercises and display them for all to see as we discuss, but instructors could try clickers or flashcards as alternatives for displaying answers simultaneously.

Since adding TBL, attendance has improved significantly in my class and is consistently near 100%. Students are less distracted during class since we transition from lecture to teamwork to discussion frequently. They seem to enjoy engaging with peers and breaking up the time we spend in class with activities.

In addition, I am confident that students have stronger ownership of their analytical skills as a result of adding TBL to my course. Developing an environment for teamwork and discussion in class seems to generate more questions from students than ever before. They are asking deeper, more thoughtful questions, and more frequently. And if we spend more time on one topic because students are interested and have extension questions, I'm happy to cut something else to accommodate that depth in understanding. Students also seem more confident in their coding skills because they are getting more hands-on practice, partnering with friends to write their code, and having more opportunities to raise issues in class instead of getting stuck with coding errors at home. Finally, I see their confidence in their new skills when they are coaching each other during TBL exercises. I am most convinced of their understanding when I am listening to the students discussing their code and interpreting results in groups. Creating the space in class to experiment has given us all more opportunities to learn from each other.

As is highlighted in TBL texts and anecdotal discussions with others who use this style, TBL is not without its challenges [1]. New materials take time to prepare, and increased discussion may take up class time once used to introduce additional topics. TBL also requires supporting students with discussion of motivation and benefits since it can be a dramatic departure from their other classroom experiences. End-of-semester reviews generally show amplified student feedback—both extremely positive and extremely negative. I highly recommend reading Sibley and Ostafichuk's text and discussing with others using this style if you're thinking about incorporating TBL into your course in the future. Sibley and Ostafichuk's text covers both the positive elements of TBL and the challenges so new practitioners are better prepared.

A student recently visited my office to ask for a recommendation as he applied for admission to graduate business analytics programs. Of course, I was happy to see him and hear about his recent activities and hopes for the future. He said that our class had made him "more skeptical of everything," ask more questions, and believe less of what he hears. For me, this feels like one step in the right direction; if students become even slightly more analytical in their thinking, then they've developed just one more professional skill that will be useful in their next adventures.

References

1. Sibley, Jim and Peter Ostafichuk. *Getting Started with Team-Based Learning*. Sterling, VA: Stylus Publishing, 2014. Print.
2. Brickell, Lt. Col. James L., Lt. Col. David B. Porter, Lt. Col. Michael F. Reynolds, and Capt. Richard D. Cosgrove. "Assigning Students to Groups for Engineering Design Projects: A Comparison of Five Methods." *Journal of Engineering Education*, 83 (3), 1994, pp. 259–262.
3. Hake, Richard R. "Interactive-Engagement versus Traditional Methods: A Six-Thousand-Student Survey of Mechanics Test Data for Introductory Physics Courses." *American Journal of Physics*, 66 (1), 1998, pp. 64–74.
4. Russell, I. Jon, William D. Hendricson, and Robert J. Herbert. "Effects of Lecture Information Density on Medical Student Achievement." *Journal of Medical Education*, 59, 1984, pp. 881–889.
5. Deslauriers, Louis, Ellen Schelew, and Carl Wieman. "Improved Learning in a Large-Enrollment Physics Class." *Science*, 332, 2011, pp. 862–864.

SECTION II

Curricular and Cocurricular Assignment Design

4

Formative and Summative Assessments in Teaching Association Rules

MATT NORTH

Utah Valley University

Contents

Introduction

This chapter focuses on two elements of effective teaching in data analytics: assessment of learning and the analytic technique of Association Rules. The goal of this chapter is to help educators understand the essential role of assessment when teaching analytics, and to provide a tangible example of how to use assessment techniques to confirm that learners understand the analytics technique being taught. The concepts of formative and summative assessment are explained and illustrated and then reiterated through a hands-on tutorial, which demonstrates how to use the R statistical software package to generate and interpret Association Rules. By the end of the chapter, you should be able to explain both formative and summative assessment, how to calculate confidence and support percentages for Association Rules, and how to generate Association Rules in R. Since formative and summative assessments are a teaching and learning technique (as opposed to a data analytics technique), you should be able to demonstrate the application of such assessments in other instructional activities, whether related to analytics or not.

Active Learning

Let's make this chapter interactive. To participate, here's what you're going to need:

- A piece of paper and a pen or pencil
- 4 tablespoons (US measure) of salt
- 8 tablespoons of flour
- 3 tablespoons of water
- Food coloring (if you're really into this)
- A spreadsheet or basic calculator

Now believe it or not, these items are going to help you learn about a data mining/analytics technique called Association Rules, and importantly for this particular chapter, how to use formative and summative assessments in teaching Association Rules. We'll start with sa non-Association Rules example in order to focus on the assessment part first.

To start, combine the salt, flour, and water in a bowl and mix the ingredients together until your items start to combine uniformly. Once this happens, knead the product together until it is completely consistent in texture, something approximately the feel of PlayDoh™ or modeling clay. If it is too sticky, add a very little bit of flour until it is not sticky anymore. If it is too dry, do the same with a drop or two of water at a time. Once you reach the desired consistency, if you'd like, you can add a few drops of food coloring and knead it some more until the coloring is consistent. Congratulations! You've just made salt dough—a popular, inexpensive alternative to other molding and modeling products, popular for preschools, craft groups and rainy day kitchen table activities. Take your little ball of salt dough and set it aside for now.

Next, take your paper and writing implement and sketch out a picture of an elephant. Don't worry; you don't have to be an amazing artist. Just draw what you think an elephant should look like. If you're not sure, take a little time and look at a few pictures of elephants on the Internet, or in books or photos that may be available to you. Don't try to copy any of the images exactly, just use them to help you envision the different attributes of an elephant—body, legs, head, ears, tail, trunk—and then draw your elephant. It's OK if you need to try two or three times, this is part of the exercise. It may be helpful to show your sketch to another person and ask them for feedback.

With your sketch now completed, return to your salt dough. Your task is to mold your ball of dough into an elephant that looks as much like your sketch as possible. Take your time. If you need tools to help you (such as a rolling pin and butter knife), go ahead and use them. Imagine that once you are done, you'll present your work, both your sketch and your model, to a third party to be judged. This person will determine how well you followed the instructions you've been given and how you used those instructions to create the most realistic and accurate salt dough elephant you can.

Formative versus Summative Assessment

This exercise may seem a bit silly, but it helps to contextualize both formative and summative assessments in teaching. You started by simply receiving and following some specific instructions. Some prerequisite knowledge was assumed: you know what salt, flour, and water are, and you know how to measure in tablespoons and mix ingredients. Other knowledge was not assumed: how much of each ingredient to use, and how to combine them. This type of direct instruction is among the most common styles of teaching used (Hadi Mohammad, Nasrin, & Maryam Taleb, 2016). Direct instruction is the teaching style most learners are familiar with, and while not always the most effective, it is often the one we default to when teaching. When the subject matter is formulaic, such as teaching a recipe, it serves as an effective teaching strategy, and most learners can benefit from direct instruction, regardless of their preferred learning styles (Tomkins & Ulus, 2016).

In this example, your instructor could come to your workspace and determine whether or not you have followed instructions correctly. Keep in mind that our objective in this chapter is not for you to learn to make salt dough—the making of salt dough is a step in the lesson, but not one of our learning outcomes. It is simply part of the teaching and learning process. The instructor's evaluation of your ability to successfully make salt dough is an example of a *formative assessment.* Formative assessments are usually informal reviews and examinations of a learner's understanding of a subject or technique being taught, in order to determine the degree to which the learner is understanding and acquiring the intended knowledge (Clinchot et al., 2017). If the ultimate task of this lesson is to teach you, and other learners, how to make model elephants out of salt dough, then the formative assessment of your salt dough helps to ensure that you're on the right track early in the learning process. If you were given all of the instruction on making a salt dough elephants all at once, and then left to your own devices to make one without any intervention, feedback or help along the way; and worse, if you made your salt dough wrong at the outset, the odds of you successfully making a salt dough elephant would decrease, perhaps dramatically, to the point where you might fail completely in your elephant-making efforts. So it is with most instruction: learners are more likely to achieve comprehension if checked along the way, and when necessary, corrected or reinforced (Klute, Apthorp, Harlacher, & Reale, 2017).

Next we move to the task of sketching your elephant on paper. This step was not presented to you in the form of direct instruction. Instead, you were left somewhat on your own to figure out how to sketch an elephant. Some recommendations were given, such as searching the Internet or books for examples of elephants. This was to help you envision what an elephant looks like, in case this was difficult for reasons that may range from unfamiliarity with elephants to a lack of aptitude in transferring thoughts to paper. It was also suggested that you might ask others for help or feedback. This task is in the vein of what is called Concept Attainment, a teaching and learning technique most often associated with constructivist theory, where students learn through guided exploration that helps them to acquire new knowledge by correctly identifying examples (and nonexamples) of the topic being taught (Bruner, 1985). Rather than being told what an elephant is, or having the attributes of an elephant dictated to you, you were charged with exploring and examining the topic on your own, seeking and evaluating sources of information, and then acting upon your learning activities to complete the task. Again, this activity would be the subject of a formative assessment along your journey to making your salt dough elephant. Your instructor could review your sketch of your elephant to determine if you were able to successfully "attain" the "concept" of an elephant and commit that concept to paper. If you got stuck along the way, your instructor might have been one of the resources available to help you learn what you needed to know to sketch an elephant. Although the model of teaching changed for this task, you were still able to learn, still able to demonstrate completion of the task, and your work could still be formatively assessed.

Finally, you returned to your salt dough and fashioned an elephant, based on the sketch you created in the prior step. The teaching style for this step changed yet again. You were not given direct instruction, nor were you asked to attain some concept. Instead, you were simply given a task and then set free to complete it. A few suggestions were given, such as using some tools if they would be helpful to you (e.g. a rolling pin), but very little specific guidance or requirements were given. This style of teaching could be described by many names: experiential instruction, project-based instruction, tactile instruction, etc., but regardless of the moniker we place on it, the emphasis in this phase of the lesson is on *learning by doing* (Meloy, 2012). This is the actual task that you set out to accomplish: to mold a salt dough elephant. In this sense, the learning outcome for this example lesson might be stated as follows: "At the conclusion of this lesson, students will be able to shape a model of an elephant in salt dough." It is when we reach this point in a lesson that we shift from *formative assessment* to *summative assessment*. Summative assessment is generally formal in nature, and is intended to specifically examine the extent to which the learner has acquired the intended knowledge (Dixson & Worrell, 2016). In this exercise, your instructor would be the ultimate authority to determine whether or not you achieved the goal of shaping a salt dough elephant, and to what degree of mastery. In the instructions, it was suggested that you might show your finished product to a peer for feedback. This would still be considered formative assessment, since that person's comments could be used to make modifications before a final evaluation of your work was performed. Only the final assessment is considered summative, although instruction and assessment should usually be considered cyclical in nature, with a new cycle of instruction and formative activities to begin after a summative assessment is made (Harrison, Könings, Schuwirth, Wass, & van der Vleuten, 2017).

Association Rules

With the concepts of formative and summative assessments now illustrated, we can examine the roles these play in the teaching of Association Rules. As with any analytic technique, there is a strong possibility that some students will struggle to understand the concepts and thus, fail to attain the desired level of understanding. We will begin with some basic direct instruction on the topic.

Direct Instruction

Association Rules are an analytic technique used to determine the coincidence of two or more elements in a data set. The classic example of where Association Rules are used is in shopping basket analysis: If you buy Product A, will you also buy Product B, and if so, how likely is that association? So to illustrate more clearly: If I go to my local grocery store intent on buying milk, will I also buy cookies, and if so, how likely is that outcome?

Consider an example from the e-commerce site Amazon.com. Open a Web browser, go to Amazon's home page, and use the search box to locate the item page for a Garmin Vivosmart HR+ fitness wristband. Open the item page, and note that below product image and description, there are areas called "Frequently bought together", "Products related to this item", and "Customers who bought this item also bought". When consumers search for this particular product, the Amazon website uses Association Rules to make recommendations of other products consumers might also buy in conjunction with their new smartwatch/fitness tracker, with the obvious intention of driving additional sales on their website, and thus, increasing revenue and profit. Association Rules are most often used to make recommendations to consumers in order to encourage additional purchases.

Customers who bought this item also bought

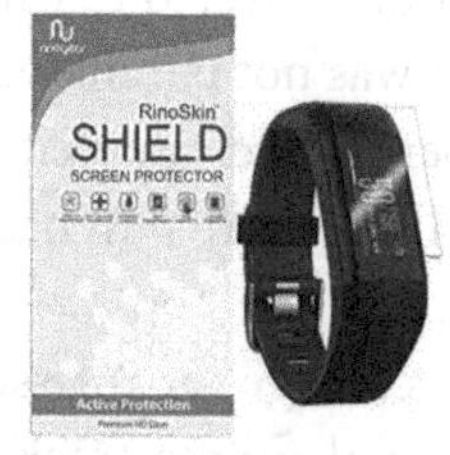

[6-PACK] RinoGear for Garmin Vivosmart HR+ Screen Protector [Active Protection] Full...
101
$4.95 prime

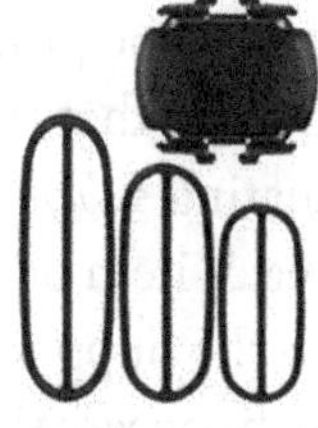

Garmin Bike Cadence Sensor
425
$33.65 prime

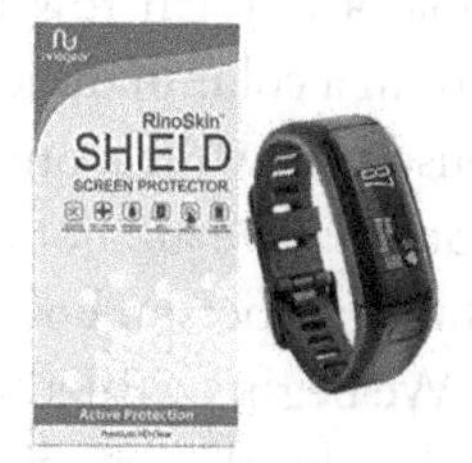

[6-PACK] RinoGear for Garmin Vivosmart HR Screen Protector [Active Protection] Full...
287
$4.95 prime

Formative Assessment #1

Now that Association Rules have been explained to you, see if you can find an example where a company other than Amazon is using this analytic technique. Find a friend or peer, and explain the concept of Association Rules using the example you found.

To illustrate what the result of Formative Assessment #1 might look like, consider Netflix. If you go to the Netflix website and search for a movie that is interesting to you, the site will also present you with other movies that are similar, or that have been watched by people who also watched the movie you searched for. Association Rules are used to identify movies to recommend to you, based on coincidence with the movie you searched for.

OK, so now we understand that Association Rules analyze data and then tell us which attributes in a data set appear to connect with one another, but how do Association Rules work? What's really happening in the background? The technique is based on frequency counts that are expressed in two fairly simple statistics: Support Percentage and Confidence Percentage. We will examine these two components of Association Rules using the example of the Garmin fitness tracker on Amazon.

Continuing Direct Instruction

We will pretend that we have access to Amazon's sales data for ten customers. We will create example data that can be used to understand how Association Rules work. Consider the mock up data displayed in Table 4.1. Each row represents an individual buyer on Amazon. A zero in a column indicates that the product was not purchased by that consumer, while a one indicates that the person did buy that product. Note that we are not evaluating *when* the items were purchased, only whether a person bought each item or not.

We begin by calculating the Support Percentage for an association. Let's take the association between the actual fitness tracker wristband (the second column from the left) and the LiQuidSkin Screen Protector (the third column from the left). We can see that eight of the ten customers in our data set bought the wristband itself. Those two who did not probably already bought the wristband somewhere other than Amazon and are now just buying accessories; or perhaps they are buying accessories for someone else who has this particular fitness tracker. In the LiQuidSkin column, we can see that five customers, IDs 47220, 50359, 94226, 70745, and 29802, purchased this item. Of those five, four customers also bought the wristband from Amazon. The association between the wristband and the screen protector *could have happened* ten times (we have ten total records in our data set), but it didn't. It actually only happened four times. Thus, our Support Percentage for the Association Rule:

Garmin Vivosmart HR+ → LiQuidSkin Screen Protector

would be 4/10, or 40%. Forty percent of the records in our data set support the rule that the wristband is associated with the LiQuidSkin

Table 4.1 Example Data of Purchase Behavior on Amazon

BUYER ID	GARMIN VIVOSMART HR+	LIQUIDSKIN SCREEN PROTECTOR	SCREEN PROTECTOR 6 PACK	CHARGING CABLE
47220	1	1	1	1
87264	1	0	0	1
15405	0	0	1	1
50359	1	1	1	0
59219	1	0	1	1
86336	1	0	0	1
94226	1	1	1	1
70745	1	1	0	0
50322	1	0	1	1
29802	0	1	0	1

screen protector. In other words, 40% of the time when a customer bought one item, he or she also bought the other item. The denominator in the Support Percentage equation is *always* the number of records in the data set, since this is always the number of times an association *could have happened.* The numerator is always the number of times that the association *actually did* occur. Customers in this data set could have purchased both the Garmin wristband *and* the LiQuidSkin screen protector from Amazon ten times, but they only did it four times out of ten, so 40% of our records *support* this association.

Next we calculate the Confidence Percentage. This is actually done twice for each possible association. The reason for this is that Association Rules uses an algorithm called "Apriori, which is a Latin term meaning "from the former." Translated into data analytics language, what this means is that we evaluate a rule in both directions, a concept we refer to as Premise → Conclusion. The premise is the product that you intended to buy and the conclusion is the product that you also bought in conjunction with the premise product. Since this could go both ways in an association between two products, we calculate the Confidence Percentage giving each item a turn to serve as the premise. So, the Confidence Percentage for the Association Rule:

Garmin Vivosmart HR+ → LiQuidSkin Screen Protector

would be 4/8, or 50%. Of the eight times a customer bought the wristband (Garmin Vivosmart HR+ serving as the premise), they also bought the LiQuidSkin screen protector four times. The denominator in the Confidence Percentage formula is *always* the number of times the premise occurred in the data set. The numerator is *always* the number of times the association occurred. So we have 50% *confidence* in this rule that is *supported* by 40% of our transaction records.

We then reverse the rule's order. The Confidence Percentage for the Association Rule:

LiQuidSkin Screen Protector → Garmin Vivosmart HR +

would be 4/5, or 80%. Of the five times the customer bought the screen protector (LiQuidSkin serving as the premise), they also bought the Garmin Vivosmart HR+ wristband four times. So, we have 80% *confidence* in this rule that is *supported* by 40% of our transaction records.

The actual meaning of these percentages is up to the analyst to determine. In this example, we have a very small sample size, so making any serious inferences about the meaning behind our findings would be unwise. But with a larger data set, say tens of thousands of purchases instead of just ten, we could begin to extract patterns of consumer behavior from the associations we see. If we had 100,000 records, and found a support percentage of just 18% for a given association, that means that that association still happened 18,000 times. If we further found confidence percentages of 65% for one premise, and 72% for the other premise, we could feel fairly satisfied that there is a true relationship between the two items, and we may then choose to cross-promote them, or offer suggestions in the same way that Amazon does.

Formative Assessment #2

Now that you've been taught about how Association Rules work and the mathematics behind them, let's see if you can calculate them on your own. Consider this association rule, using the data in Table 4.1:

$$\text{Screen ProtectorSixPack} \rightarrow \text{Charging Cable}$$

Calculate the Support Percentage and both Confidence Percentages for this rule, and then explain your findings to a peer.

The solution to this exercise would look like this:

$$\text{Screen Protector Six Pack} \rightarrow \text{Charging Cable : Support Percentage} = 5/10 = 50\%$$

$$\text{Screen Protector Six Pack} \rightarrow \text{Charging Cable : Confidence Percentage} = 5/6 = 83.3\%$$

$$\text{Charging Cable} \rightarrow \text{Screen Protector Six Pack : Support Percentage} = 5/8 = 62.5\%$$

- The association between these two products *could have* occurred ten times. It *did* occur five times. Five times out of ten records yields 50% of the records supporting this association.

- The Screen Protector Six Pack was sold six times. The rule occurred in five out those six transactions, so we have 83.3% confidence that when the premise is the Six Pack, the customer will also buy the Charging Cable.
- The Charging Cable was sold eight times. The rule occurred in five out of those eight transactions, so we have 62.5% confidence that when the premise is the Charging Cable, the customer will also buy the Screen Protector Six Pack.

Going Big

Now that you've had some instruction, and two formative opportunities to test out how well you understand Association Rules, it's time to face the facts: ten records are far too small of a data set to yield any meaningful results. Even if it weren't, think about the complexity of evaluating combinations of products. For example, what if the premise is that a customer comes to Amazon to buy both the Garmin fitness tracker *and* a six pack of screen protectors? Assuming a two-product premise like that, what is the Confidence Percentage in a conclusion of Charging Cable? As we add more columns of possible products that could be associated, more rows of customer purchases, and more possible combinations for associations, the complexity of generating and evaluating Association Rules becomes hopelessly complex if we try to do the frequency counts and calculations manually. This is where tools come in. This example will use the R statistical package with RStudio, to demonstrate how a simple, free software tool can help you quickly and easily generate Association Rules of varying degrees of complexity.

Using R

The R statistical package is a free, open source analytics tool. RStudio is a development environment, also freely available, that makes working with data in R a little bit easier. R can be downloaded from www.r-project.org/, and RStudio can be downloaded from www.rstudio.com/.

If you wish to work alongside this example, download both software products and then install the statistical package first, then install

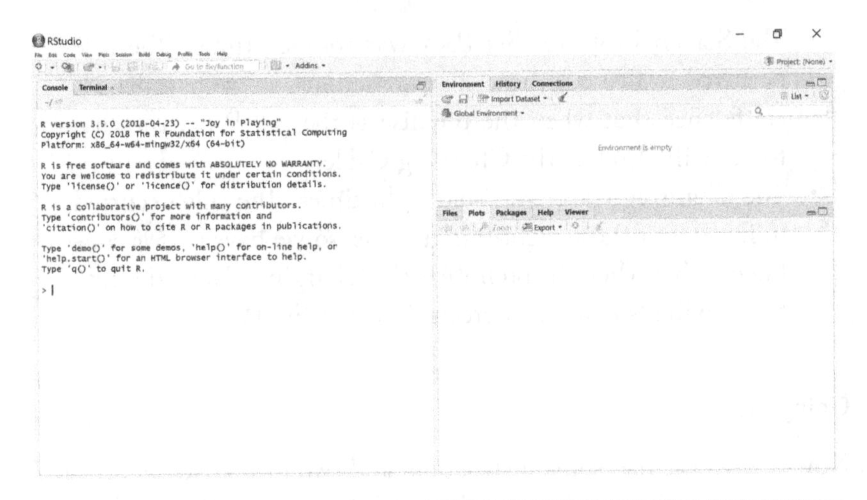

Figure 4.1 RStudio running on top of the R statistical package on Microsoft Windows 10.

RStudio. Once you have these installed you can launch RStudio. Your environment will look something like Figure 4.1.

To create Association Rules in R, we will need to add a package to R called "arules." This is done in the RStudio application menu by clicking Tools, then Install Packages (Figure 4.2).

Type "arules," without the double quotes, and click Install to add this package to your R environment (Figure 4.3).

RStudio will show some feedback indicating that the "arules" package has been added. We're now ready to examine our data set at a much larger scale. For this example, the same Garmin fitness tracker example data set is used, but it has been expanded to include 5,848 rows of mock up data. It has been saved as a comma delimited (CSV) file (Figure 4.4).

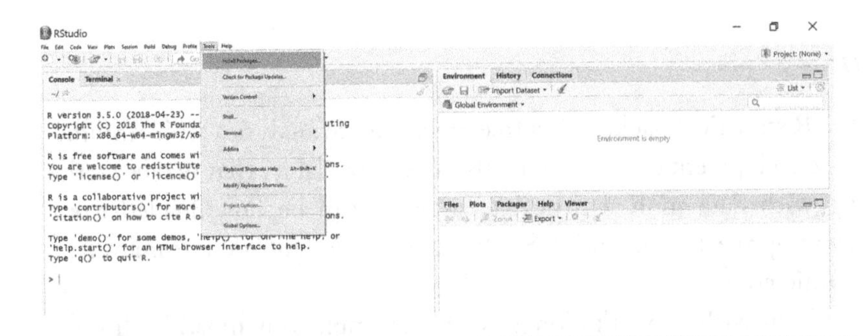

Figure 4.2 Install packages in RStudio.

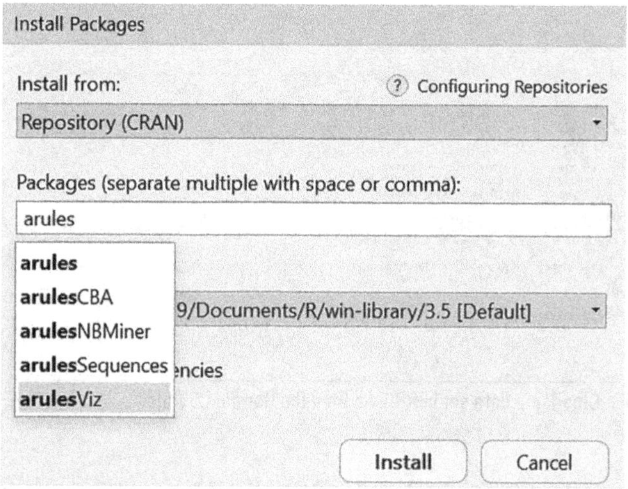

Install Packages

Install from: Configuring Repositories

Repository (CRAN)

Packages (separate multiple with space or comma):

arules

arules
arulesCBA
arulesNBMiner
arulesSequences
arulesViz

9/Documents/R/win-library/3.5 [Default]

Install Cancel

Figure 4.3 Adding the "arules" package to R.

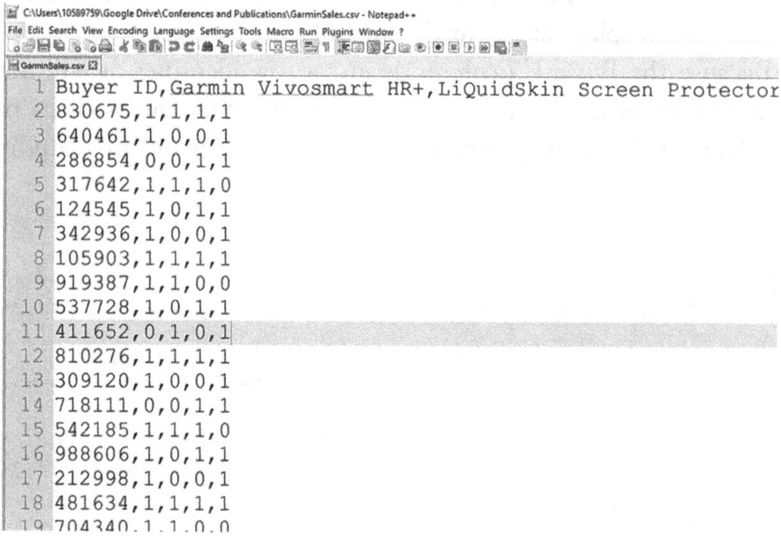

C:\Users\10589759\Google Drive\Conferences and Publications\GarminSales.csv - Notepad++

File Edit Search View Encoding Language Settings Tools Macro Run Plugins Window ?

GarminSales.csv

```
1 Buyer ID,Garmin Vivosmart HR+,LiQuidSkin Screen Protector
2 830675,1,1,1,1
3 640461,1,0,0,1
4 286854,0,0,1,1
5 317642,1,1,1,0
6 124545,1,0,1,1
7 342936,1,0,0,1
8 105903,1,1,1,1
9 919387,1,1,0,0
10 537728,1,0,1,1
11 411652,0,1,0,1
12 810276,1,1,1,1
13 309120,1,0,0,1
14 718111,0,0,1,1
15 542185,1,1,1,0
16 988606,1,0,1,1
17 212998,1,0,0,1
18 481634,1,1,1,1
19 704340,1,1,0,0
```

Figure 4.4 A partial screenshot of the expanded example data set in CSV format.

We will begin by connecting to the data set in RStudio (Figure 4.5). This is done by issuing the command (note that R is case-sensitive): GarminData <- read.csv(file.choose(), header=T, colClasses='factor').

After submitting the read.csv command in RStudio, browse to the location on disk where the CSV file is saved. Opening the file creates

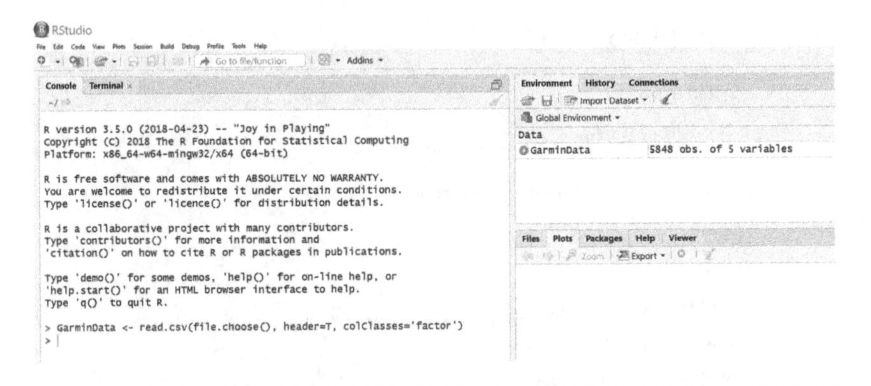

Figure 4.5 Creating a data set in RStudio from the Garmin CSV file.

a data set named GarminData in RStudio. Figure 4.5 shows that the read.csv statement was entered and executed, and that the data set is now stored in RStudio. Clicking on the data set (GarminData, showing 5,848 observations and 5 variables) will open a view of the data in a table-style display (shown in Figure 4.6).

Because the Buyer.ID column is just a row identifier, it will not be useful in seeking associations between the different products in our data set. We can remove it by issuing a command in RStudio

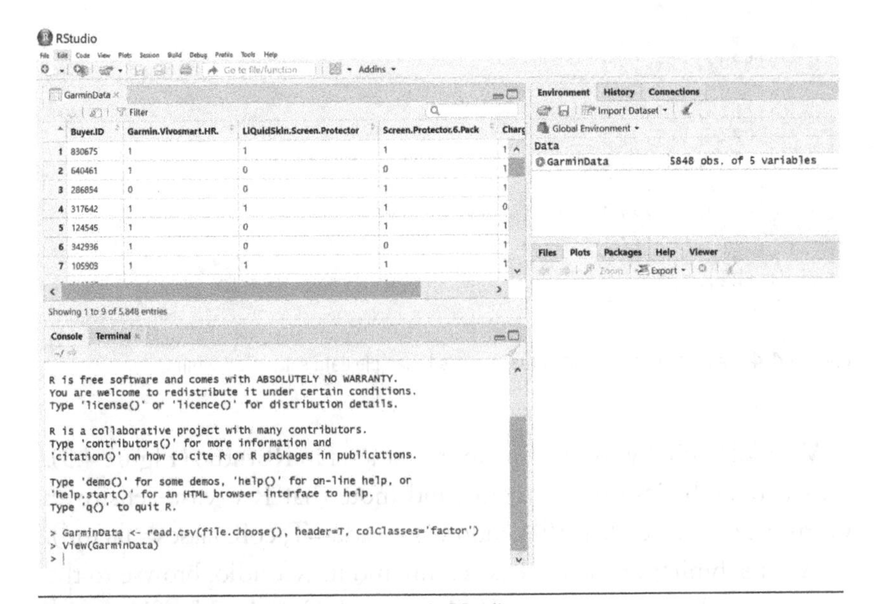

Figure 4.6 Click a data set to view the data in RStudio.

that reduces our data set to only the second through fifth columns: GarminData <- GarminData[2:5].

This command replaces the GarminData object in RStudio with a new version that only contains the product columns. You will see that when you issue this command, the view of your data in the upper left quadrant of your RStudio window is refreshed to only display columns two through five, as shown in Figure 4.7.

Next, we need to change the expression of purchasing behavior in our data a little bit. The "arules" package in R views a value of zero (0) as data, since zero is a numeric value. In preparing this purchasing data, we coded 0 to mean the absence of a purchase, so in order for R to read the data as either a purchase, or not a purchase, we need to replace all zero values in our data set with the code NA (Figure 4.8). This is easily done by issuing this command: GarminData[GarminData==0] <- NA.

We can now have R apply the "arules" package to our data, and generate Association Rules. Invoke the "arules" package by issuing the command: library(arules), then generate rules by issuing the command: GarminRules <- apriori(GarminData, parameter=list(minlen=2, supp=0.2, conf=0.5)).

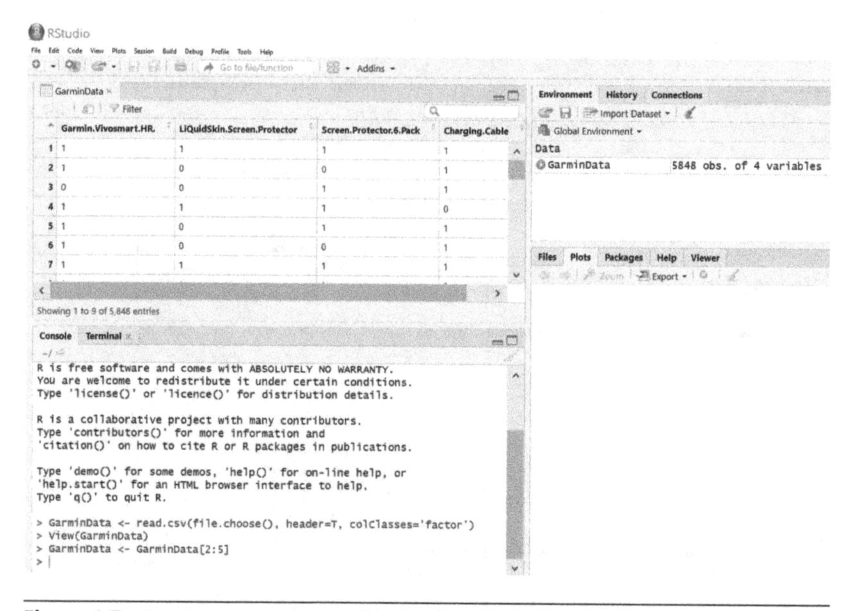

Figure 4.7 A reduced data set called GarminData, with the Buyer.ID column eliminated.

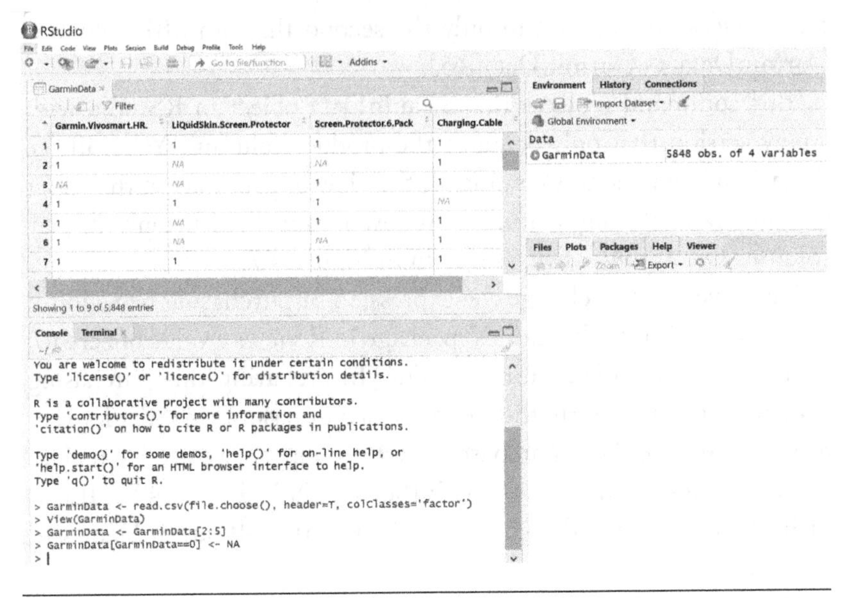

Figure 4.8 Values of 0 replaced with *NA* in R.

Figure 4.9 shows that the "arules" library was successfully loaded, and that a set of Association Rules was generated and placed into a data object in R called "GarminRules." Note that for this example, we

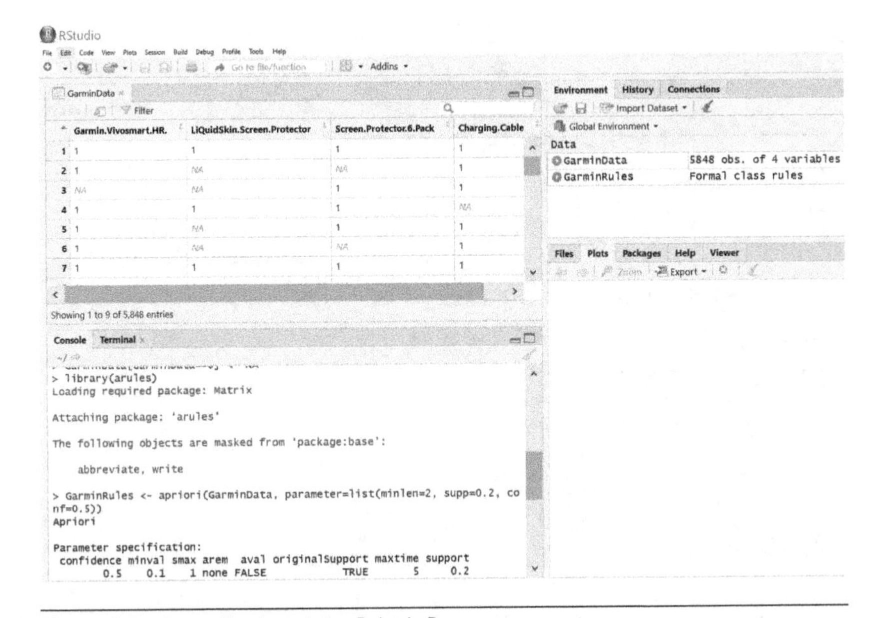

Figure 4.9 Generating Association Rules in R.

have given R some parameters to be used in identifying and creating Association Rules:

- minlen=2: This parameter means that a rule must have at least two columns (attributes) in it. This is the smallest association rule that can be created, since you need to have at least one premise and one conclusion in order to have an association.
- supp=0.2: This parameter requires that any associations R finds must have a support percent of at least 20%.
- conf=0.5: This parameter requires that any associations R finds must have a confidence percent of at least 50%.

Not pictured in Figure 4.9 is some additional information about the Apriori algorithm and the frequency patterns R was able to identify. If you are following along in R as you read this chapter, you will see this information displayed in the Console tab in the bottom left quadrant of your RStudio application window. It will look like Figure 4.10, and you should see that R was able to find and create 25 Association Rules.

Now that the rules have been generated and stored in R in the data object named "GarminRules," the easiest way to view those rules in a table format is to issue: GarminRulesTable <- as(GarminRules, 'data.frame').

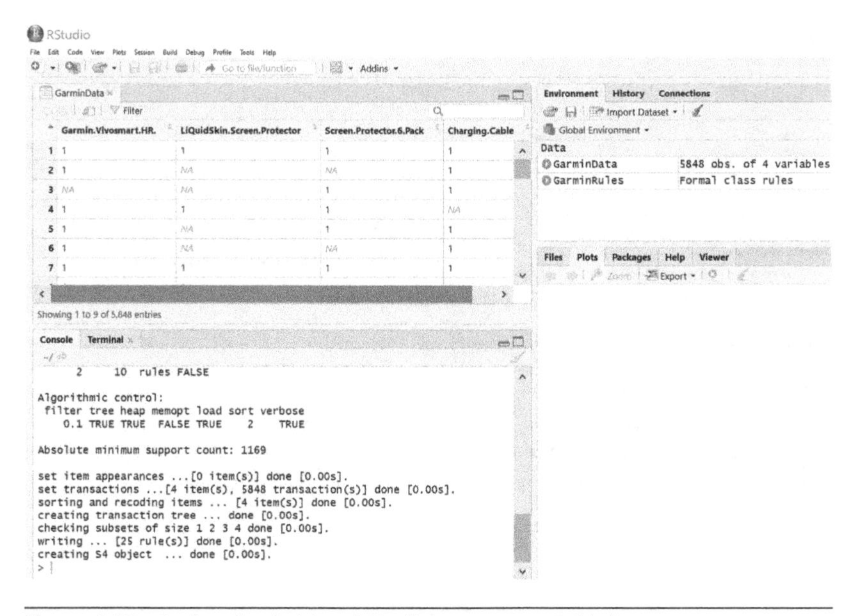

Figure 4.10 Confirmation that Association Rules were generated in R.

This will create a data frame in R named GarminRulesTable, and when clicked in RStudio, will open up the Association Rules for view, as shown in Figure 4.11.

The scroll bar on the right of the table viewing pane in RStudio will allow you to look at all 25 rules R was able to find. Notice that if you click on any of the column headers in the GarminRulesTable view, you can sort by each column. Clicking twice will sort the table in reverse order. If you click twice on the support column header, you will find that the association rule Charging.Cable → Garmin. Vivosmart.HR (rule number 10) has a support percent of 60.1% and a confidence percent of over 75.5%. This is the strongest rule in our data set, and indicates that when a person buys the charging cable, 60% of the time they also buy the fitness wristband, and that we can have 75.5% confidence in this rule. This gives strong data-based evidence that it would be wise to cross promote these two items on an e-commerce website such as Amazon.com.

Formative Assessment

Identify two more rules from the chapter's example exercise, and explain them to someone.

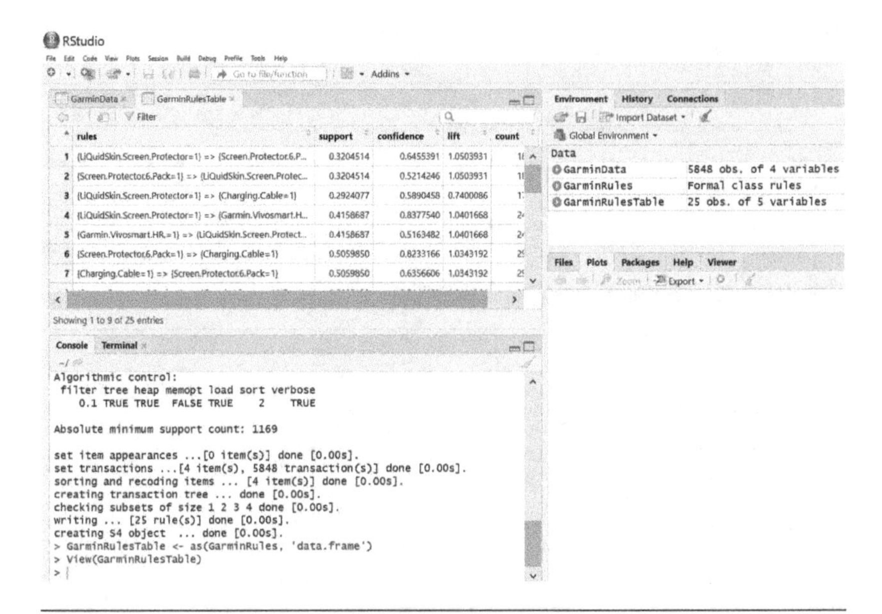

Figure 4.11 Viewing Association Rules in a named R data frame.

An example solution to this formative assessment might be

- Rule number 14: LiQuid.Screen.Protector → Screen.Protector.6.Pack. In our data, 32% of our customers bought both of these products together; and, when a person came with the intent to purchase the LiQuid Screen Protector, 100% of the time they also bought the Screen Protector 6 Pack.
- Rule number 18: Garmin.Vivosmart.HR, LiQuid.Screen.Protector → Charging.Cable. This rule is interesting because it has two premises, which lead to one conclusion. In our data, 21% of our customers bought all three of these products together; and, when a person came with the intent to purchase the Garmin fitness wristband *and* LiQuid screen protector for it, 51% of the time they also bought the Charging Cable. This is our weakest rule, given the minimum confidence and support percentages we set as parameters, yet even as our weakest rule, it still shows strong association among the products in our data set.

Summative Assessment

It is now time to see if you have learned how to create and interpret Association Rules on your own. Download and examine that data file labeled MarketBasketAnalysis.csv. Using RStudio and the example steps and code in this chapter, generate Association Rules for the products purchased as represented in the data set. Set your minimum support percentage to 15% (0.15) and your minimum confidence percentage to 50% (0.5). Don't forget to remove the Receipt ID and to take zero values out of consideration. Put your rules into a data frame so that you can view them in a table. Examine and interpret your rules. Explain what you find to a peer.

If you need help, here are the R codes you will need to issue in order to accomplish this summative assessment correctly:

- library(arules) (only necessary if you have not already invoked this library in your current R session)
- MarketData <- read.csv(file.choose(), header=T, colClasses='factor')

- MarketData <- MarketData[2:12]
- MarketData[MarketData==0] <- NA
- MarketRules <- apriori(MarketData, parameter=list(minlen=2, supp=0.15, conf=0.5))
- MarketRulesTable <- as(MarketRules, 'data.frame')

Conclusion

This chapter has illustrated how to use the teaching and learning techniques of formative and summative assessments to teach new skills in data analytics using Association Rules. Association Rules are a method of finding items (columns, or attributes) in a data set that frequently coincide with one another. This technique is useful with applications in marketing and sales but could also be applied to find any patterns in data that indicate that two or more observable phenomena often happen together. For example, medical professionals could use this technique to mine data to see if two or more illness indicators are frequently associated. Manufacturers could use it to see if two or more flaw indicators are often found together.

By applying formative and summative assessments when teaching the association rule analytic technique, you can determine whether or not learners are achieving the desired level of understanding and learning outcome. Formative assessments serve as checkpoints along the way to help students demonstrate and confirm their knowledge of foundational concepts on their way to mastery of the overall topic, which is assessed through a summative experience. Once students demonstrate satisfactory competence on the summative assessment, such as generating Association Rules on a new, unfamiliar data set and correctly explaining any identified rules, they can then use their new knowledge to solve problems or as a foundation to learn other analytic techniques.

References

Bruner, J. (1985). Models of the Learner. *Educational Researcher, 6*, 5.

Clinchot, M., Ngai, C., Huie, R., Talanquer, V., Lambertz, J., Banks, G., Weinrich, M., Lewis, R., Pelletier, P., & Sevian, H. (2017). Better Formative Assessment. *Science Teacher, 84*(3), 69–75.

Dixson, D. D., & Worrell, F. C. (2016). Formative and Summative Assessment in the Classroom. *Theory into Practice, 55*(2), 153–159.

Hadi Mohammad, P., Nasrin, Z., & Maryam Taleb, D. (2016). Teachers' Knowledge of Modern Teaching Methods and its Relationship with Student Learning Styles. *International Journal of Advanced Biotechnology and Research, 7*, 1148–1158.

Harrison, C., Könings, K., Schuwirth, L., Wass, V., & van der Vleuten, C. (2017). Changing the Culture of Assessment: The Dominance of the Summative Assessment Paradigm. *BMC Medical Education, 17*(1), 73.

Klute, M., Apthorp, H., Harlacher, J., & Reale, M. (2017). Formative Assessment and Elementary School Student Academic Achievement: A Review of the Evidence. *Regional Educational Laboratory Central, National Center for Education Evaluation and Regional Assistance*, 2017–259.

Meloy, J. M. (2012). *Twenty-first Century Learning by Doing*. Rotterdam, Netherlands: Sense Publishers.

Tomkins, L., & Ulus, E. (2016). 'Oh, was that "Experiential Learning"?!' Spaces, Synergies and Surprises with Kolb's Learning Cycle. *Management Learning, 47*(2), 158–178.

5

The Necessity of Teaching Computer Simulation within Data Analytics Programs

VIRGINIA M. MIORI

Saint Joseph's University

Contents

Analytics techniques may be categorized as descriptive, prescriptive, and predictive. Descriptive tools are used in the preliminary stage of data processing. Summaries of historical data are generated in order to create information and create data for further analysis. Descriptive methods include Web analytics, traditional descriptive statistical techniques, fitting of statistical distributions, and data visualization. Predictive techniques allow the analyst to make predictions about unknown future events such as product demand, economic growth, and facility utilization. These predictions are made using data mining, statistical analysis, machine learning, and artificial intelligence.

Prescriptive analytics are used to find the best course of action for a given situation and rely upon both descriptive and predictive analytics.

Computer simulation is among the most powerful tools used in prescriptive analytics and is a highly underrepresented tool in data analytics programs. Teaching of structured simulation engages students while enhancing their modeling skills by creating computer representations of physical systems. The physical systems come from settings including, but not limited to manufacturing, process, inventory, service, or education.

All simulations begin with the generation of a baseline analysis, which breaks down a system into component parts, identifying the behavior and performance of each part. After completion and validation of the baseline, scenario identification and processing are performed. Output from the baseline and the scenarios are gathered, analyzed, and tested for significant differences. Students are able to assess the effectiveness of proposed scenarios in achieving system improvements in both time and cost savings. In fact, the complete simulation approach provides a blue print for success in assessing virtually any business problem.

In this chapter, we will focus on *Discrete Event System Simulation* (DESS) which refers to the modeling of a system as a discrete sequence of events in time. DESS is appropriate for any type of operation that may be easily viewed as discrete events, though not for continuous processes like extrusion operations.

Process Mapping

Process mapping, or flow charting, is the first step to writing simulations. It also marks the first significant challenge for students. They must be introduced to a process map as a symbolic representation of a physical process that requires a detailed understanding of the process or system. Well-defined system boundaries are also critical to effective mapping. Specific examples and personally relevant examples are required to help students transition into process thinking.

Process order and process boundaries can be reinforced by asking students to model processes in their own lives: cooking meals, doing laundry, study routines, etc. Then, they may be introduced to broader process examples.

Example 1: Describe the process for a local restaurant with a desire to serve the student population.

Example 2: Describe the process for a relevant manufacturer (Samsung, Apple, Dell, etc.) that begins with a standard set of raw materials and ends with well-defined finished products.

There is also significant value in examining the alternative type of process, one with multiple and disparate manufacturing operations. These would be inappropriate for a single process map, but a network of process maps could be generated to reflect this increased complexity.

Process mapping uses standard symbols to represent the flow of any system or process. It is not necessary to learn every possible symbol. Too much focus on an extended set of symbols tends to distract student attention away from the process breakdown and toward the more controllable question of which symbols to use. Six basic symbols are in fact enough to effectively create a process map to be used as a first step in simulation modeling.

The terminator symbol represents the entry and exit points for the simulation. Terminators are shaped like rounded rectangles and are typically labeled "Start" and "End." Sample terminators are presented in Figure 5.1.

Arrows are used to represent the flow between blocks. They further define the predecessor/successor relationships and are typically single-directional. Rectangles represent individual actions within a process; they may be manufacturing, service, or other actions within a process. Diamonds indicate decision points that allow for two possible choices. A series of diamonds is required to represent more complex decisions. Labeled circles or connector symbols are used to extend process maps to multiple pages and inverted isosceles triangles are used to represent queues. All of these symbols are shown in Table 5.1.

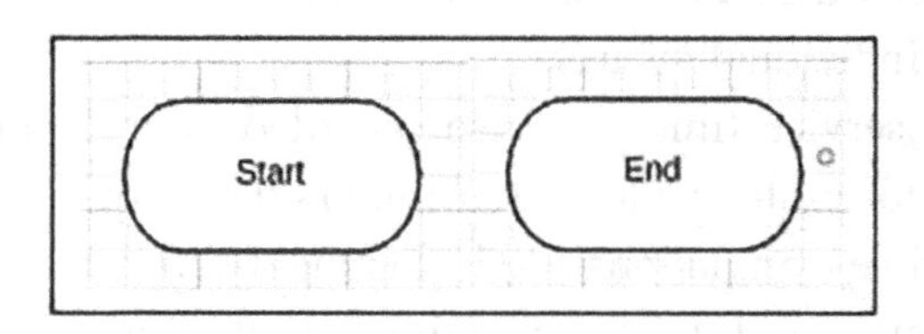

Figure 5.1 Terminator blocks.

Table 5.1 Process Mapping Symbols

Symbol	Description
→	Arrows (flow indicators)
□	Action (process) block
◇	Decision block
Ⓐ	Page connector
▽	Queue (waiting time)

Process maps may be created using various software products. Students may purchase software such as Microsoft Visio, or they may use free software. The free software options do limit the functions available, but typically provide enough capability to fulfill student needs. Links to software options are available in Appendix 5B.

We now introduce the process example to be used throughout the chapter.

Process Example: Consider a back with three tellers. Each teller has its own line (queue). Customers arrivals to the bank are recorded based on the moment they enter the bank, time between arrivals has been calculated as well. The bank is open for 8 h each day from 9:00 am to 5:00 pm. Customers entering the bank before 5:00 pm will be served, but no customers will be allowed to enter the bank after 5:00 pm. The bank would like to determine whether it is more efficient to have a single queue that disperses to all three tellers rather than the tellers having individual queues.

Customer service times are also recorded upon completion of service for each customer. Customers may require processing of multiple transactions and individual transaction processing time may vary. Table 5.2 contains arrival times, and Table 5.3 contains service times.

Table 5.2 Customer Interarrival Times

2.80	6.02	24.99	6.57	2.99	1.12	0.06	4.87	6.82	26.00
12.15	0.82	8.01	5.07	13.78	2.32	11.25	2.25	1.69	4.74
9.69	0.69	23.26	6.58	8.00	1.40	0.15	1.60	1.07	15.53
24.09	2.40	2.00	16.50	18.16	2.82	0.11	6.14	5.25	0.48
2.00	18.36	6.40	2.71	4.32	0.87	2.52	4.17	7.30	0.90
1.88	46.16	6.76	20.48	5.81	6.93	9.84	7.91	8.09	0.03
24.24	0.03	1.53	4.53	4.62	0.82	0.67	5.94	1.52	1.19
10.36	10.86	5.81	4.68	11.71	15.54	17.87	3.77	0.89	4.51
1.46	4.14	17.37	5.13	2.18	5.47	0.52	22.84	1.01	2.95
25.96	2.82	2.24	13.05	1.73	2.18	6.32	0.99	5.04	2.06

Table 5.3 Customer Service Times

8.03	13.85	11.62	10.51	8.30	14.57	12.00	16.59	7.09	7.92
13.82	8.61	4.37	13.43	11.05	11.41	8.31	5.51	7.01	6.21
11.10	8.24	17.28	7.11	7.50	10.81	12.35	9.35	12.93	12.71
4.82	12.35	8.84	9.57	13.08	9.50	9.29	10.49	11.97	12.88
17.23	14.41	11.57	12.31	10.75	11.99	13.21	10.81	5.95	13.13
8.82	14.79	7.23	14.06	8.46	11.15	9.08	11.34	11.12	12.25
11.28	7.95	7.94	8.13	2.65	9.70	8.23	8.87	8.44	9.40
13.06	14.08	7.86	13.14	10.94	11.09	4.74	8.47	14.47	11.81
11.50	12.72	6.47	11.17	16.96	14.70	8.57	12.13	6.55	10.50
10.69	13.53	13.09	8.51	6.95	12.77	6.28	7.83	6.70	10.52

The process map for the current queue (baseline) configuration bank is shown in Figure 5.2, while the process map for the proposed single-queue configuration in shown in Figure 5.3. These process maps are created with horizontal flows, but students may alternatively employ a vertical process flow.

Variations in process map configurations are much more likely as processes become more complex. In the case of the bank example, we would expect students to closely match Figures 5.2 and 5.3 when creating their process maps. It is also helpful to ask students to provide a verbal description as documentation.

Process Example: In the baseline process, after entering the bank, customers must determine which queue is the shortest. Since decision blocks have only two possible outcomes, a series of two decision blocks is used to find the shortest queue. The first decision block asks whether Queue 1 is shortest, if

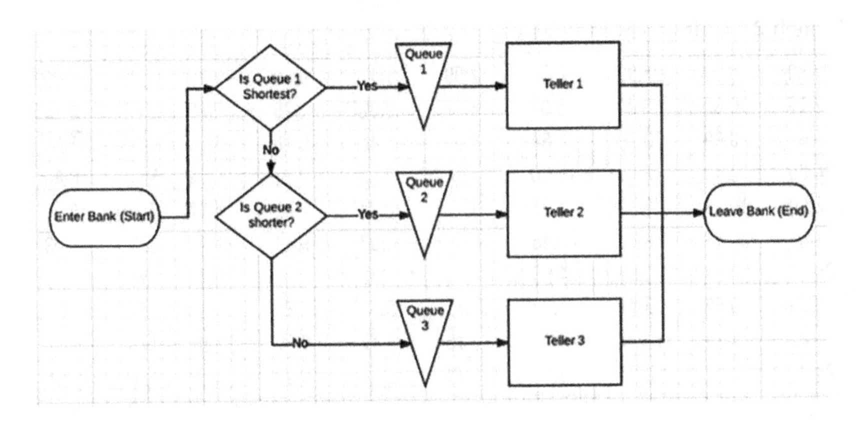

Figure 5.2 Bank process map (current).

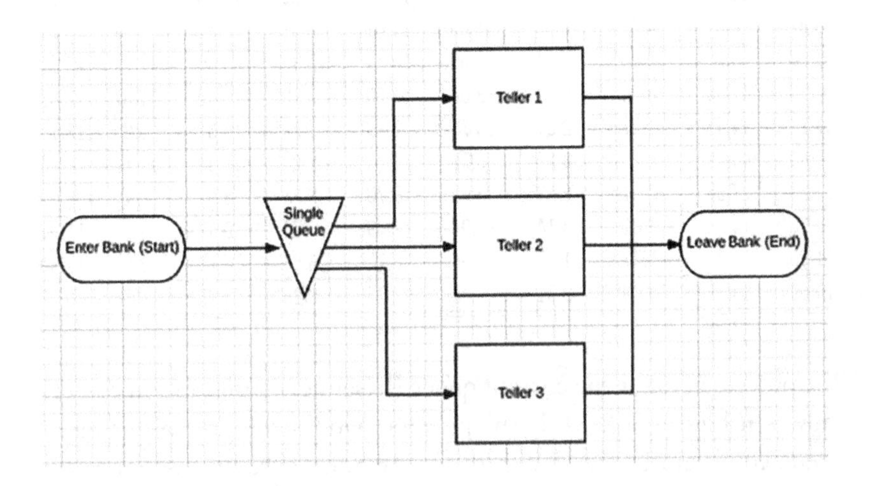

Figure 5.3 Bank process map (proposed).

the answer is yes, the customer processes to Queue 1. If the answer is no, the second decision block asks whether Queue 2 is shorter. If Queue 2 is shorter, the customer processes to Queue 2, if not, the customer processes to Queue 3. Once in the queues, customers wait until the Teller is available. They approach the teller, receive the requested service and leave the bank.

In the single-queue process, customers enter the bank and join a single queue. They are served in order of arrival to the bank, receive the requested service, and depart from the bank.

Fitting Statistical Distributions

After completing the process map, effective simulation depends on the student's ability to accurately represent historical data (service times, interarrival times, etc.). The arrival and activity processing times must be characterized and noted on the process maps. On occasion, circumstances of a system may result in static (constant) interarrival times and processing times. It is, however, more common for these times to be dynamic in nature, therefore requiring analysis to best describe the data. Fitting a statistical distribution to historical data is the ideal approach to characterizing the data. Specialized simulation software products readily use statistical distributions; even Microsoft Excel makes easy use of distributions.

It is often the case that simulation software comes packaged with software designed to fit statistical distributions. One such example is ExtendSim® software that comes packaged with Stat::Fit® software. Stat::Fit easily finds the best fitting distribution for any data set. JMP® statistical software also provides this capability through its continuous fit function. Links to statistical fitting options are provided in Appendix 5B.

Fitting of distributions in this way may be unfamiliar to even seasoned analytics students. Before teaching the fitting techniques, a review of typical statistical distributions may be helpful. In particular, students should have familiarity with the uniform (real) distribution, the normal distribution, the exponential distribution (including the relationship to and conversion from the Poisson distribution), empirical distributions, and the triangular distribution. Understanding the nature of these distributions will provide a sufficient base for students to extrapolate their understanding to more obscure distributions.

Process Example: Figures 5.3 and 5.4 present results of distribution fitting within JMP. Figure 5.4 provides a summary of the interarrival times to the bank. In order to find the best fit, we ask JMP to fit all distributions and select the distribution with the smallest error value. JMP uses the Akaike Information Criteria (AIC) to measure the quality of the fit. In this case, the exponential distribution was the best fit for the interarrival data, having the smallest value of the AIC. Though we

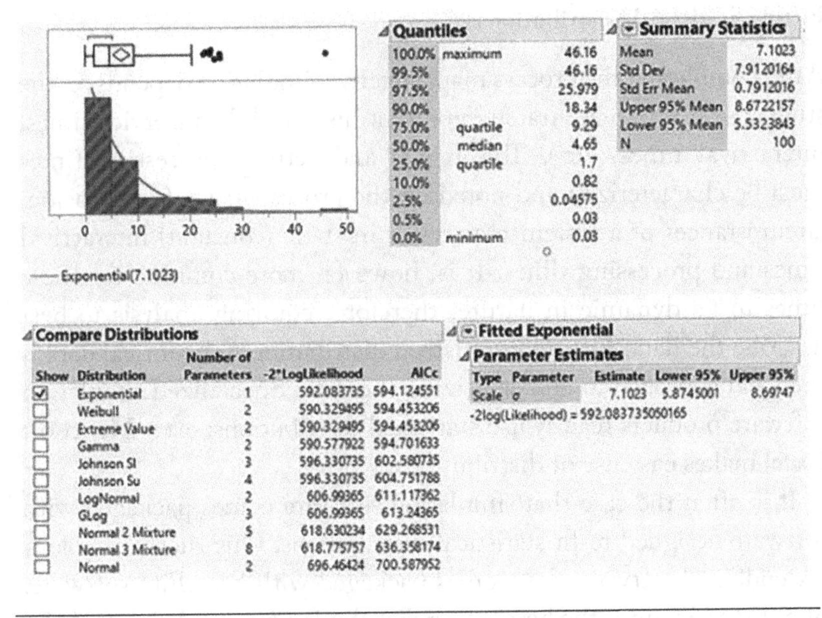

Figure 5.4 JMP output fitting interarrival data.

are not displaying the alternative distribution histograms, we are able to see the fit and AIC values for all other candidate distributions.

The service time data fit is provided in Figure 5.5. In this case, the Weibull distribution was the best fit. Despite the Weibull being an unfamiliar distribution, simulation software like ExtendSim is very robust and allows for specification of all distributions. When faced with a need to work within statistical distributions known to students, and noting that the difference in the AIC between the Weibull and Normal distributions is very small, we may instead elect to use the Normal distribution results.

Based on this analysis, the interarrival times for the bank example should be described as exponentially distributed with a mean of 7.102 min between arrivals. The service times may be described as either Weibull distributions with $\alpha = 11.517$ and $\beta = 3.905$, or a Normal distribution (see output in Appendix 5A) with $\mu = 10.424$ and $\sigma = 2.990$.

With the process maps in hand and statistical distributions identified, students are ready to tackle basic queueing theory concepts.

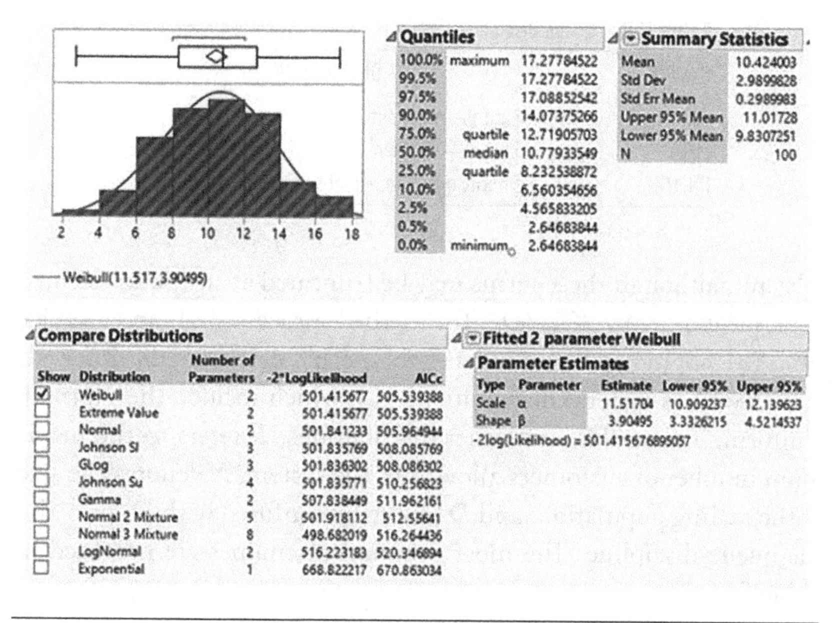

Figure 5.5 JMP output fitting service time data.

Basic Queuing Theory

It's important to discuss basic queuing theory but equally as important to avoid losing students in the theoretical mathematics. It is valuable to understand that queues are categorized in specific ways and that these categories are embedded in our process maps and simulation models. It is also important for students to grasp that certain statistical distributions are extremely common in queueing theory.

The most basic queuing system, and therefore the most commonly used, is the *M/M/1* model. The first position (***M***/*M/1*) denotes arrivals to the system. The *M* indicates that the interarrival times are exponentially distributed. The second position (*M/**M**/1*) denotes the service in the system. The *M* again indicates that service times are exponentially distributed. The third position (*M/M/**1***) denotes the number of servers. Note that the M is more specifically referring to the Markovian or memoryless property of the exponential distribution.

A generalized parallel multiserver model with exponential interarrivals and exponential service is the *M/M/c* model where *c* is the number of servers. This extension is also very commonly seen in practice.

Students benefit from the standard approach and the complete designation of a queuing process. Three additional terms are

Table 5.4 Queue Disciplines

SYMBOL	DESCRIPTION
FIFO	First in—First out
LIFO	Last in—First out
PNPN	Priority queue (preemptive and not preemptive)

relevant, although these terms may be truncated at any point deemed appropriate, *A/S/c/K/N/D*. The arrival process (*A*) and service time (*S*) notation allow students to specify deterministic intervals (D) as well as general distributions (*G*), which include the Normal, Uniform, Triangular, and other distributions. *K* refers to the maximum number of customers allowed in the system, *N* denotes the size of the calling population, and *D* (not to be confused with D) provides the queue discipline. The most common disciplines are provided in Table 5.4.

Process Example: The queuing model for the bank example is specified as *M/G/3*. *G* is specifically found to be the Normal distribution.

Simulation Using ExtendSim

Though simulation may be accomplished using Excel, it is quite burdensome for students to take this path. Using software developed specifically for simulation results in streamlined and effective learning as well as more applicable outcomes. Several programs are available and commonly used in academics, including Arena®, AtRisk® and Crystal Ball®. We employ the ExtendSim software in this chapter. Links for all of these software products are provided in Appendix 5B.

ExtendSim software is visual software that allows users to select blocks when building their simulation model. Each block has at its core, a collection of programs and subprograms written to emulate process behavior. The software may be used easily by novices, but also maintains complex functional ability to support the needs of expert modelers making it ideal for students.

The modeling blocks are distributed into libraries based on function. It's useful to encourage students to expose the library block lists in order to provide a scaffold for model building (Figure 5.6).

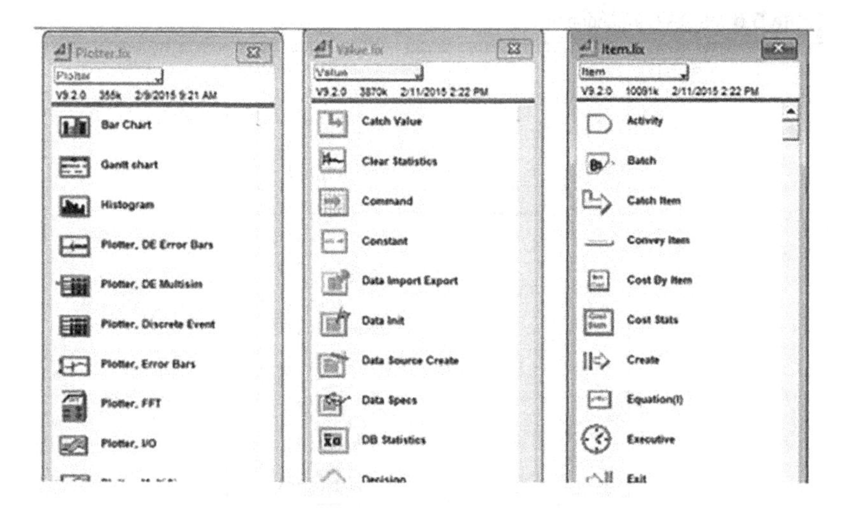

Figure 5.6 Sample ExtendSim blocks. (ExtendSim blocks copyright © 1987–2017 Imagine That, Inc. ExtendSim is a registered trademark of Imagine That Incorporated in the United States and/or other countries.)

Students should become familiar with the types of blocks stored in each library. The most commonly used libraries are the Item Library, that contains all blocks that relate to the physical processing of items through the simulation, the Value Library that holds blocks associated with the flow of information, the Plotter Library that is used to produce plots and charts, and the Utilities Library that contains blocks that support customization of the simulation interface. Additional libraries should be set aside until students become more experienced users.

Using the process map, students can generate a nearly one-to-one relationship between the process mapping blocks and the simulation blocks.

> Process Example: The corresponding blocks for the bank example are presented in Table 5.5. Students' attention should particularly be drawn to the idea that a single *Select Item Out* block replaces the cascading decision blocks in the process map. Additional useful ExtendSim blocks are noted in this table.

The best approach to simulation for anyone is to begin modeling with the simplest structure, this is particularly important for students.

Table 5.5 Block Correspondence

Flowchart	ExtendSim
 Enter Bank (Start)	 *Create Block*
 Leave Bank (End)	 *Exit Block*
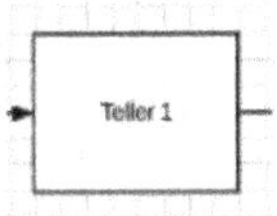 Teller 1	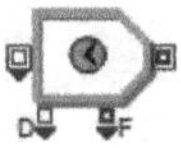 *Activity Block*
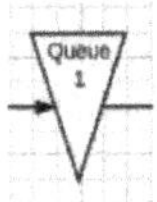 Queue 1	 *Queue Block*
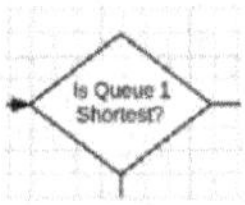 Is Queue 1 Shortest?	 *Select Item Out Block*
	 *Executive Block*
	Max & Min Block
	Statistics Block

They often attempt to work more quickly and find the code very difficult to debug. Debugging can be a daunting process; in this way, we always have working components and may more effectively debug any new logic. Once the simple structure is proven to work, complexities may be added.

> Process Example: The additional logic to manage the end of day at the bank will be postponed until the logic for the internal workings of the bank has been completed.

Before beginning a simulation, modelers must determine the length of time that the simulation will run, the number of simulation runs necessary and the appropriate global time units.

> Process Example: The bank will be open for an 8 h day (480 min). Ultimately, we would like to run the simulation 1,000 times, and generate average performance statistics, to ensure a representative solution.

When first creating a simulation, it should be run once. After debugging and finalizing a simulation, students should plan on running the simulation hundreds or even thousands of time in order to accurately represent the typical state of a system. In particular, students should be made aware that ExtendSim software requires an executive block to be placed in every simulation, without it, models will not run.

> Process Example: The simulation setup for the bank is provided in Figure 5.7, while the complete simulation model is presented in Figure 5.8. Note that each block has been labeled to assist in understanding the flow.

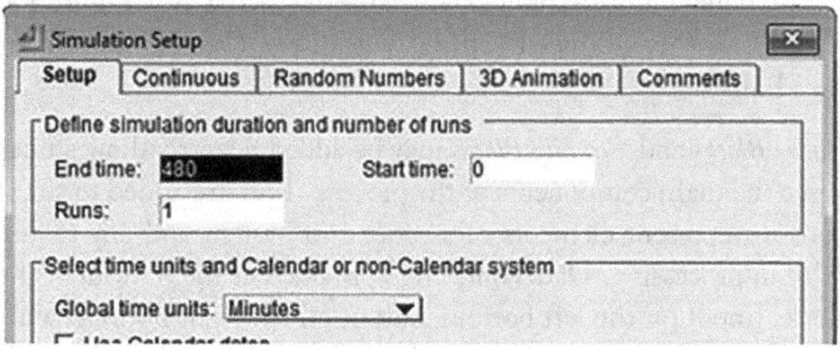

Figure 5.7 Simulation setup dialog box.

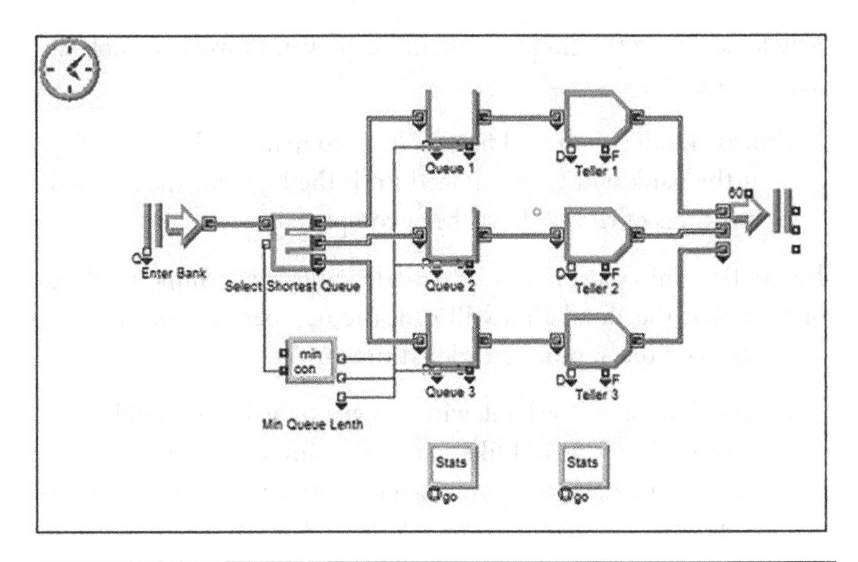

Figure 5.8 Bank example simulation model.

Preliminary Blocks

Each time a student adds a simulation block to a model, the associated parameters must be set within the block properties. Simulation languages offer many options in specifying timing for blocks including constants, single statistical distributions and multiple statistical distributions. Many more complex options and parameters are available and may be explored in more advanced courses. Following the major logic flow of the process map helps students make a smooth transition into building a simulation model. The *Create Block* models arrivals to the system.

Process Example: We found that the interarrival time was exponentially distributed with a mean of 7.102 min. Figure 5.9 shows the parameter settings within the *Create Block* (Enter Bank).

Queue Blocks and *Activity Blocks* may be added next, to allow students to see the main components of the process. They are added to simulations to represent elements of a process or system and any required delay in processing. Data connectors provided at the bottom of these blocks (input on the left bottom, output on the right bottom) will be important in helping students establish the information flow. All data connectors may be dragged and dropped (down) to expose all possible

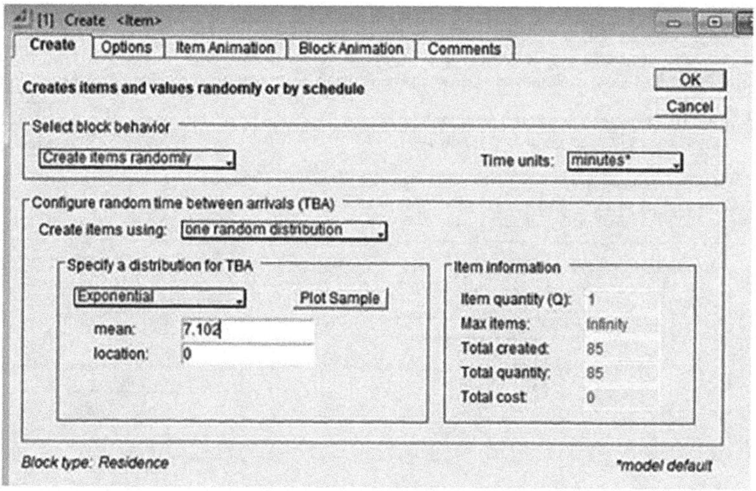

Figure 5.9 Interarrival time specification in create block.

input and output values. The established standard for information flow is the use of single lines, item flows are represented with double lines. The same standard applies to input and output connectors; single boxes indicate data flow and double boxes indicate item flow.

Process Example: Students must represent the current state of the bank as one queue per teller. Queues are confirmed to be sorted queues using the FIFO discipline (first-in, first-out) in Figure 5.10. The service times for all tellers are entered as

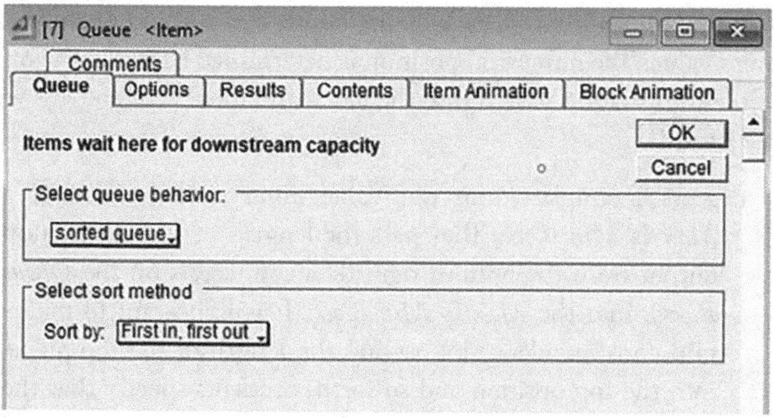

Figure 5.10 Queue characteristics.

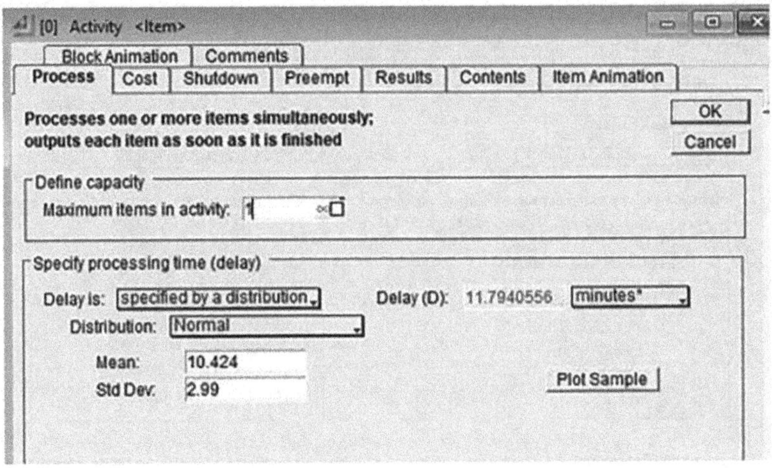

Figure 5.11 Service time specification in activity blocks.

Normal distributions with a mean of 10.424 and a standard deviation of 2.99 in Figure 5.11. If data had been collected separately for each teller, we would have been able to specify unique distributions for each teller.

When faced with multiple queues, the selection of the shortest queue is not automatic in ExtendSim. Students must therefore build decision logic using blocks from the Value Library to assist the simulation items in finding the shortest queue. The *Min & Max Block* is particularly useful to students when finding the shortest queue. The block can accept a large number of input values and has two output values: the minimum value and the numerical position of the minimum value. The numerical position is determined based on the order of the information passed into the block and numbering begins with the value 0.

Process Example: From the Value library, students select a *Max & Min Block*. They pass the lengths of the queues (top output from the bottom right data connectors on the *Queue Block*), into the *Max & Min Block*. It is important to maintain correspondence by passing the length of the top queue into the top position and so forth. Students specify that the "minimum value" is sought in Figure 5.12 and pass the associated position to the *Select Item Out* (decision) *Block*.

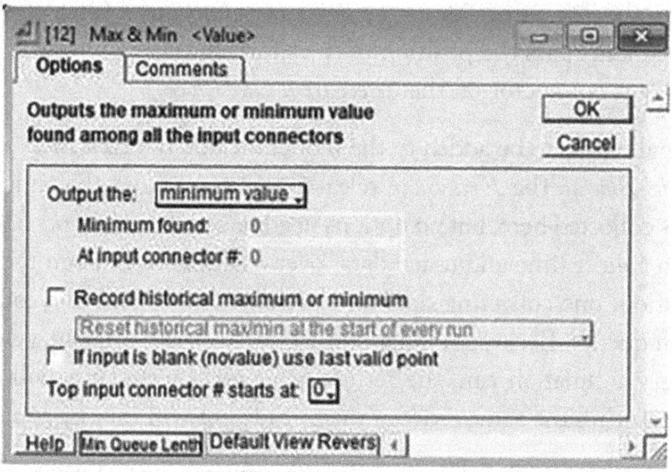

Figure 5.12 Max & Min block option settings.

The *Select Item Out Block* allows complex decisions to be made, eliminating the need for successive decision blocks. A number of options are available to users in specifying the method for determining the correct output path. These include the option to select the output path based on the use of a "select connector" as shown in Figure 5.13. The output paths are also numbered starting with 0. In the case of a tie, the left-most (in simulations created vertically) or top-most (in simulations created horizontally) queue will be selected.

Process Example: Students must first choose the "select connector" method before the corresponding input connector appears

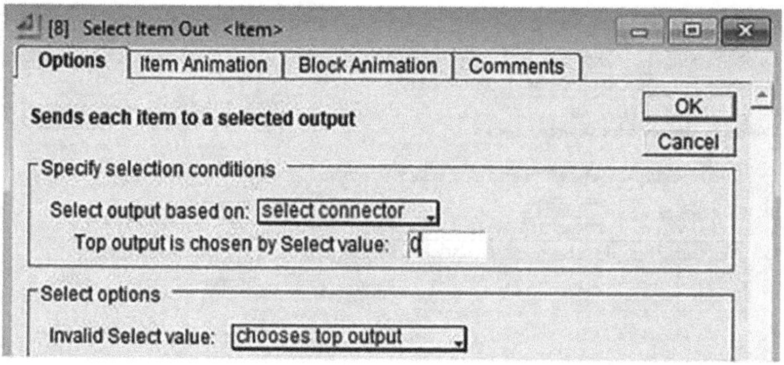

Figure 5.13 Select Item Out options.

on the left-hand-side of the *Select Item Out Block.* Once visible, students pass the position of the shortest queue directly to the select connector on the *Select Item Out Block.*

The final blocks to be added to the model include the *Exit Block* and the *Statistics Blocks.* The *Exit Block* releases the items from the simulation. Data is collected here, but no options need be set. The *Statistics Blocks* are used to collect time and usage data, over multiple runs, from the entire simulation, one collecting data on activities (tellers) and one collecting data on queues. Because simulations use pseudo random data, generated for every simulation run, the results generated by students will always remain within the same order of magnitude but will vary by run.

Process Example: Figure 5.14 (activities) and Figure 5.15 (queues) both show results of a single run of the bank simulation. In the case of multiple runs, an additional row of data would be output for each run.

The importance of documentation and block labeling becomes clear when examining the statistics. Student-assigned block labels are used to reference each row of results.

Process Example: The recorded values for the tellers include arrivals and departures for each teller, teller utilization, average number of customers served (Ave. Length) and service time (Wait) among other values. The first teller has the highest utilization because it always wins the tie when all queues are the same length.

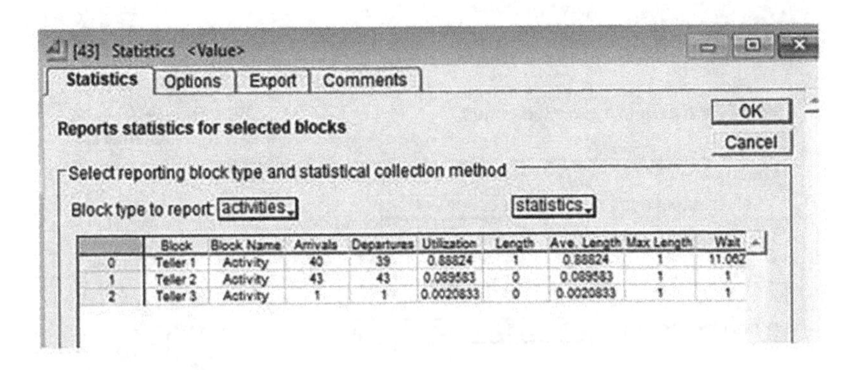

	Block	Block Name	Arrivals	Departures	Utilization	Length	Ave. Length	Max Length	Wait
0	Teller 1	Activity	40	39	0.88824	1	0.88824	1	11.062
1	Teller 2	Activity	43	43	0.089583	0	0.089583	1	1
2	Teller 3	Activity	1	1	0.0020833	0	0.0020833	1	1

Figure 5.14 Activity (teller) statistics.

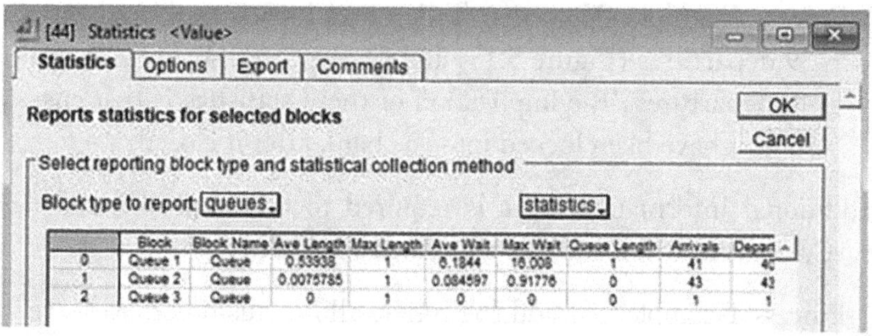

[44] Statistics <Value>

Statistics | Options | Export | Comments

Reports statistics for selected blocks

OK

Cancel

Select reporting block type and statistical collection method

Block type to report: queues

statistics

	Block	Block Name	Ave Length	Max Length	Ave Wait	Max Wait	Queue Length	Arrivals	Depart
0	Queue 1	Queue	0.53938	1	6.1844	16.008	1	41	40
1	Queue 2	Queue	0.0075785	1	0.084597	0.91776	0	43	43
2	Queue 3	Queue	0	1	0	0	0	1	1

Figure 5.15 Queue statistics.

The queue statistics are also referenced by block label. The average queue length, waiting time, arrivals, and departures are some of the wvalues provided. The queue statistics dialog is shown in Figure 5.15.

Debugging

Any time programming is taught, one of the most critical concepts to enforce with students is that debugging takes two forms. The first and simpler form is debugging for syntax. Were commands and blocks entered correctly? The second and more insidious issue occurs when program logic is not correct. In other words, the code does what the student has requested, but the student's request was incorrect.

Both forms of debugging must be completed on the basic flow before adding any complexities to a simulation model. Students are encouraged to *Show 2d Animation* (from the *Run* menu) to track the flow of items through their simulation. Using animation, which slows the simulation significantly, allows students to discern whether a block is processing correctly (syntax error) and to watch each item process through the model to check the system logic against the logic of the simulation.

Adding Complexity

Once the basic flow has been debugged, students may then turn to the addition of complexities in the simulation. Complexities may include closing time logic, resource allocation, activity capacity, and conditional process logic.

Process Example: Note that Teller 1 had 40 arrivals and only 39 departures (Figure 5.15) and Queue 1 has 41 arrivals and 40 departures. The implication of these statistics is that customers have been locked into the bank after it closed.

Additional information logic is required to allow all customers to leave a process at the end of a simulated work day.

Process Example: To add logic that allows customers to leave the bank at the end of their transaction, students must add a *Simulation Variable Block*, a *Constant Block*, and a *Decision Block* from the Value Library. Another *Select Item Out Block* and an *Exit Block* from the Item Library (Figure 5.16) are also required.

As seen in Figure 5.17, a *Constant Block* set to 480 (min) represents an 8 h day. Students will find that the *Simulation Variable Block* may be set to a number of different simulation variables associated with time, runs, dates, etc. Of particular importance is the "Current time" that stores the time elapsed in the simulation. All of the information in this block is collected automatically during the simulation processing.

Process Example: Students should set the *Constant Block* to 480 min, and the *Simulation Variable Block* should be set to store the current simulation time. Students must extend the

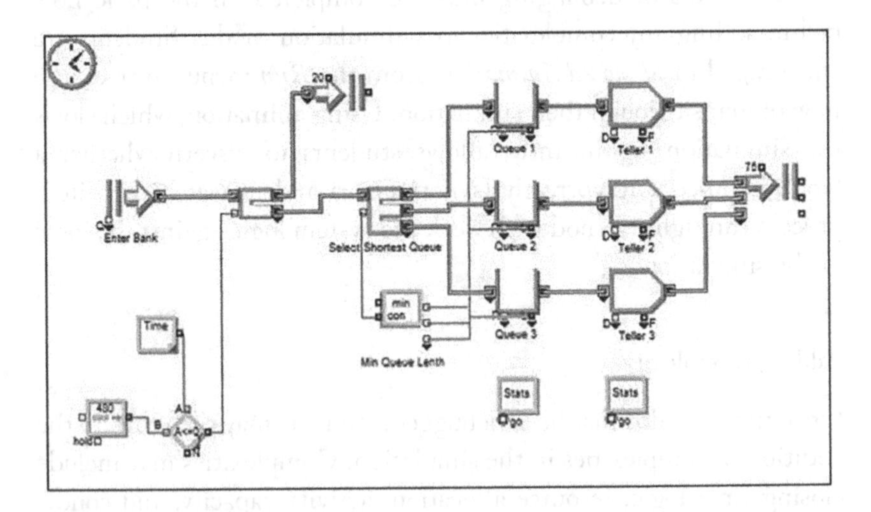

Figure 5.16 Revised simulation.

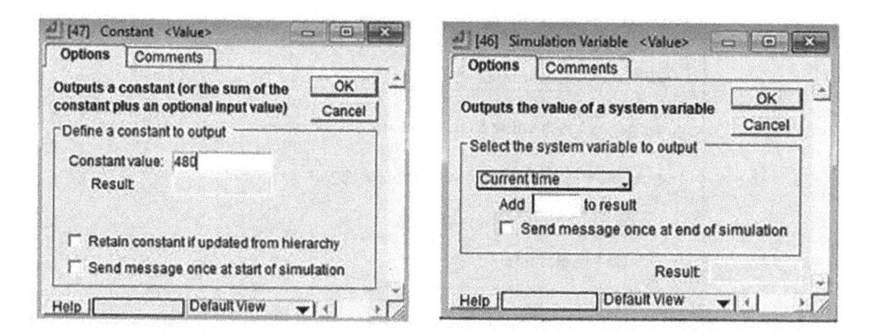

Figure 5.17 Setting constant and simulation variable options.

End Time to allow the simulation to run long enough to allow all customers to leave (Figure 5.18) but cannot allow customers to enter the bank after 5:00 pm (minute 480).

The *Decision Block* within the Value library takes the traditional diamond shape. It allows two inputs, A and B, and allows the user to determine the nature of the decision based on a selected condition. The Y (Yes) connector will output a 1 if the condition evaluates to True, and 0 if it evaluates to False. The N (No) connector will output a 0 if the condition evaluates to True and a 1 if the condition evaluates to False. Although the decision block operates in this traditional manner, students should be encouraged to allow unconventional use of the decision outputs.

Process Example: Figure 5.19 shows the decision is set to evaluate whether A (current simulation time) is less than or equal to B (closing time). Rather than using both decision outputs, the student may use only the Y connector. This value may be

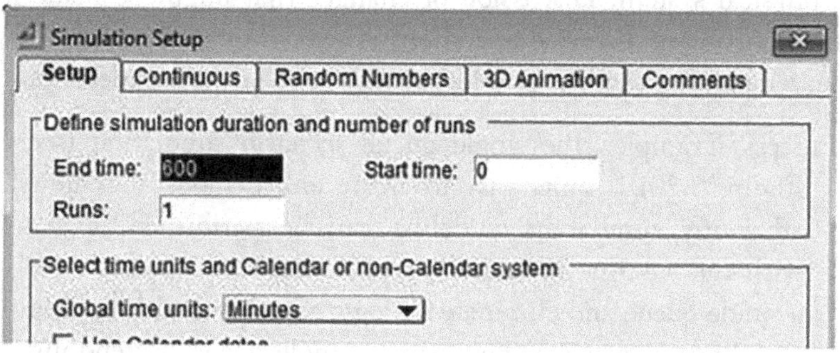

Figure 5.18 Extend the simulation time.

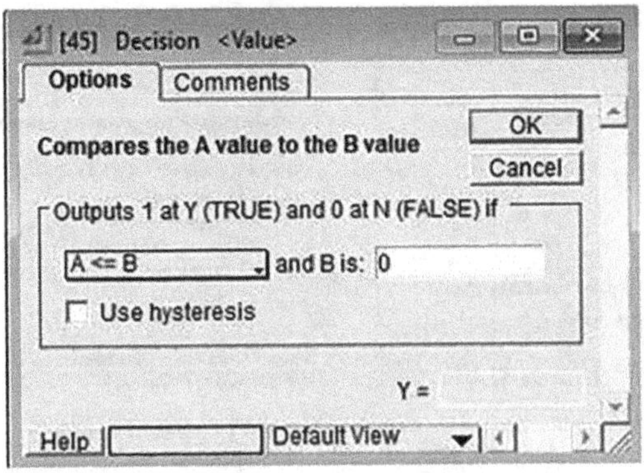

Figure 5.19 Decision block options.

passed to the *Select Item Out Block* (recall the top output path is number 0). A value of 0 from the Y connector can be used to route the customer directly to an *Exit Block,* since it indicates that the bank's doors are closed for the day. A value of 1 allows the customer to process through the bank. The extended simulation time allows customers to finish being served, while keeping new customers from entering the bank after closing time.

Once all complications have been coded, the students should save this simulation as a baseline for analysis. The original question posed in the process example was whether it would be more efficient for customers to process through a single queue for all three tellers. Students should treat this question as the basis for a what-if scenario analysis. Creation of scenarios with ultimate comparison to the baseline model exemplifies the true power of simulation. Modelers may evaluate the effectiveness of what-if scenarios without making changes to the physical system. Discussion of changes that might be made in manufacturing and service operations is critical to assisting students in understanding that the physical changes can be quite expensive.

Process Example: The single queue scenario simulation (see Figure 5.20) is quite easy to create and demonstrates how efficiently simulation modeling can be performed. Begin with the baseline simulation, replace the three queues with a single queue and eliminate all logic associated with finding the shortest queue. When using a single queue, ExtendSim automatically selects the first available activity.

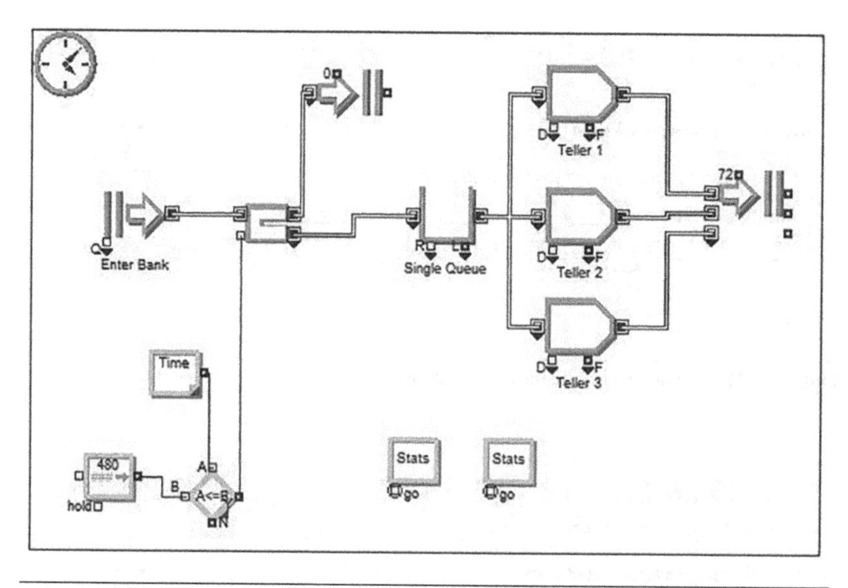

Figure 5.20 Single-queue scenario simulation.

Juxtaposing the simplicity of changing a simulation model to the complexity of changing a physical process is quite powerful to students. Presentation of an actual case study further supports the argument for teaching simulation.

Output Analysis

Statistical analysis, in particular hypothesis testing, is necessary in generating final validation of simulation models. The baseline data outcomes must be compared to actual operational results. If these results have been found to be statistically equivalent, students may conclude that the baseline is an accurate representation of the physical process. If the baseline is not shown to be statistically equivalent to the physical process, it indicates that the simulation model is either incomplete, incorrect, or exceeds the boundaries of the process. This forces students back to the drawing board to rebuild the baseline and reassess process data.

Statistically equivalent results allow students to further compare what-if scenarios to the baseline. When students identify statistically significant differences between scenarios, they may further conclude that these differences reflect actual differences in operational configurations.

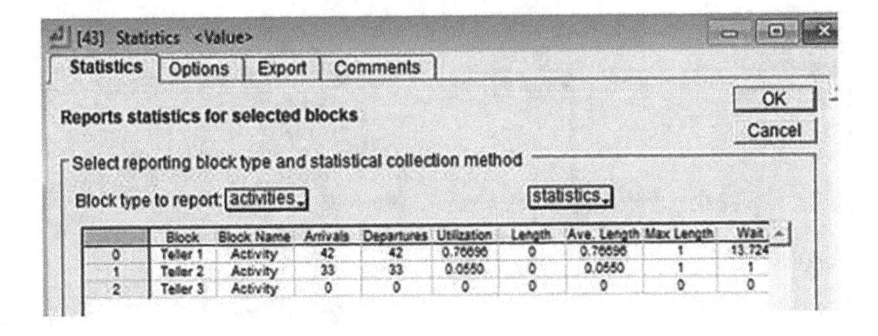

[43] Statistics <Value>

Statistics | Options | Export | Comments

OK

Cancel

Reports statistics for selected blocks

Select reporting block type and statistical collection method

Block type to report: activities

statistics

	Block	Block Name	Arrivals	Departures	Utilization	Length	Ave. Length	Max Length	Wait
0	Teller 1	Activity	42	42	0.76696	0	0.76696	1	13.724
1	Teller 2	Activity	33	33	0.0550	0	0.0550	1	1
2	Teller 3	Activity	0	0	0	0	0	0	0

Figure 5.21 Single-queue scenario activity statistics.

[44] Statistics <Value>

Statistics | Options | Export | Comments

OK

Cancel

Reports statistics for selected blocks

Select reporting block type and statistical collection method

Block type to report: queues

statistics

	Block	Block Name	Ave Length	Max Length	Ave Wait	Max Wait	Queue Length	Arrivals	Depart
0	Queue 1	Queue	0.37302	1	5.3289	12.949	0	42	42
1	Queue 2	Queue	0.0059137	1	0.10752	0.9936	0	33	33
2	Queue 3	Queue	0	0	0	0	0	0	0

Figure 5.22 Single-queue scenario queue statistics.

Process Example: In examining the results of a single run, we can see that the average customer wait has decreased from 6.18 to 5.33 min (Figures 5.21 and 5.22). This apparent improvement must be tested by running each simulation 1,000 times, with the calculation of cumulative statistics.

After 1,000 runs of the base case and 100 runs of the single queue scenario, the overall average waiting times in the queues are presented in Table 5.6. The overall average waiting time in the baseline was 1.706 min, while the average waiting time in the single queue scenario was only 0.002 min. This is an impressive reduction but must be validated statistically.

A two-sample T test is an appropriate test for two scenarios. We perform an upper tailed test to allow a conclusion of whether the average waiting time for the baseline is greater than the average waiting time for the single-queue scenario. The hypotheses and alpha level are

Table 5.6 Accumulated Statistics over 1,000 Runs

SCENARIO—QUEUE	AVERAGE WAITING TIME (MIN)	STD (MIN)
Baseline—Queue 1	5.053	0.694
Baseline—Queue 2	0.770	0.036
Baseline—Queue 3	0.004	0.033
Baseline—Overall average	1.706	2.401
Single-queue scenario—Single queue	0.002	0.004

```
Two-Sample T-Test and CI

Sample     N      Mean    StDev  SE Mean
1       3000      1.71     2.40    0.044
2       1000   0.00200  0.00400  0.00013


Difference = mu (1) - mu (2)
Estimate for difference:  1.70400
95% lower bound for difference:  1.63187
T-Test of difference = 0 (vs >): T-Value = 38.87  P-Value = 0.000  DF = 2999
```

Figure 5.23 Two sample *T* test results.

presented below, and the statistical output is presented in Figure 5.23. With a test statistics of $T = 38.87$ and a p-value of 0.000, we resoundingly reject the null hypothesis and conclude that the average waiting time for customers at the bank is reduced when a single line is implemented.

$$H_0 : \mu_1 = \mu_2$$

$$H_1 : \mu_1 > \mu_2$$

$$\alpha = 0.05$$

Concluding Remarks

The inclusion of simulation within any data analytics curriculum helps students to integrate the seemingly disparate tools that we teach. The ability to see how analytic techniques leverage each other and fit together yields the additional educational benefit of enhancing students modeling and critical thinking skills. Simulation facilitates the inclusion of complex business problems into undergraduate and graduate curriculums. Data analytics programs and program reputations are enhanced, all for the improved use of analytics in the workplace.

More advanced statistical tests may also be employed when examining simulation outputs. These include goodness-of-fit tests, run tests, and other nonparametric methods. We may employ other multivariate analysis tools such as ANOVA and various types of regression when examining detailed outputs.

ExtendSim and other simulation software tools are available to universities at little cost. ExtendSim provides a free network license along with a deeply discounted student version of the software. See Appendix 5B for relevant links.

Appendix 5A: Output for Normal Distribution

See Figures 5A.1 and 5A.2.

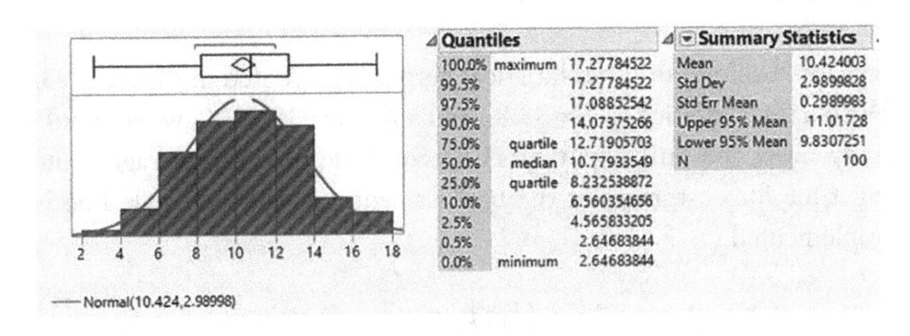

Quantiles

100.0%	maximum	17.27784522
99.5%		17.27784522
97.5%		17.08852542
90.0%		14.07375266
75.0%	quartile	12.71905703
50.0%	median	10.77933549
25.0%	quartile	8.232538872
10.0%		6.560354656
2.5%		4.565833205
0.5%		2.64683844
0.0%	minimum	2.64683844

Summary Statistics

Mean	10.424003
Std Dev	2.9899828
Std Err Mean	0.2989983
Upper 95% Mean	11.01728
Lower 95% Mean	9.8307251
N	100

Figure 5A.1 Descriptive summary of fit.

Compare Distributions

Show	Distribution	Number of Parameters	-2*LogLikelihood	AICc
☐	Weibull	2	501.415677	505.539388
☐	Extreme Value	2	501.415677	505.539388
☑	Normal	2	501.841233	505.964944
☐	Johnson Sl	3	501.835769	508.085769
☐	GLog	3	501.836302	508.086302
☐	Johnson Su	4	501.835771	510.256823
☐	Gamma	2	507.838449	511.962161
☐	Normal 2 Mixture	5	501.918112	512.55641
☐	Normal 3 Mixture	8	498.682019	516.264436
☐	LogNormal	2	516.223183	520.346894
☐	Exponential	1	668.822217	670.863034

Fitted Normal

Parameter Estimates

Type	Parameter	Estimate	Lower 95%	Upper 95%
Location	μ	10.424003	9.8307251	11.01728
Dispersion	σ	2.9899828	2.6252254	3.473389

-2log(Likelihood) = 501.841233023142

Figure 5A.2 Fitting data to the normal distribution.

Appendix 5B: Resource List

Process Mapping

Business Enterprise Mapping (www.businessmapping.com/blog/best-business-process-mapping-software/)

Zak, R. (2018). 9 best free alternatives to Microsoft Visio, *Maketecheasier*, Retrieved from www.maketecheasier.com/5-best-free-alternatives-to-microsoft-visio/

Statistical Fitting

Mathwave: Data Analysis and Simulations (www.mathwave.com/articles/fit-distributions-excel.html)

Mathworks – fitdist, Fit probability distribution object to data (www.mathworks.com/help/stats/fitdist.html?s_tid=gn_loc_drop)

Simulation Software

Arena® Simulation Software (www.arenasimulation.com/academic)

Palisade Academic Software (www.palisade.com/academic/)

Oracle Crystal Ball (www.oracle.com/us/products/applications/crystalball/overview/index.html)

ExtendSim: Power Tools for Simulation (www.extendsim.com/academic)

ExtendSim Adopter Program (www.extendsim.com/adopter)

6

Using Games to Create a Common Experience for Students

STEPHEN PENN

Harrisburg University of Science and Technology

Contents

Many difficulties exist in teaching analytics to college-level students. Students come from different backgrounds and finding a particular domain to discuss that excites everyone is very difficult. Finding data sets that are interesting and complex enough for serious analytics work usually means spending time explaining the structure, interpretation, and value of the given data set. If the instructor picks a particular algorithm to teach, such as association analysis to explore proportions and relationships, the research question is set after the method is chosen, instead of the research question driving the data, which drives the method. Consequently, the chosen algorithm seems arbitrary to the students. A method of generating student interest, having the students develop an understanding of the data, and having all students share the same level of understanding of the real-world process of producing the data is critical to effectively teaching any given analytical technique. Having a common understanding, a similar level of expertise, and an interest in generating questions to be answered with analytics is the Xanadu of teaching analytics.

My attempt at solving this problem is to play a game in the classroom. In playing a game, individuals and teams must make decisions based on limited information within predefined rules. Games create a challenging, interesting, fun, and (especially) shared experience. This chapter discusses the selection of a game, how to introduce students to a game, options in playing a game, examples of games played, how to debrief the students after the game, how to reference the game played in future lectures, examples of assignments based on game play, and other benefits.

Playing games in the classroom has several benefits, but the selection, introduction, execution, and follow-up need to be planned. The games discussed in this chapter are not logical games, such as prisoner's dilemma, nor are they video games. They are published board games that have competition, lots of data, and many opportunities for decision-making. Games that I've used in the classroom have covered managing a factory, investing in the stock market, and recreating the Battle of Hastings in the year 1066. These games create shared experiences for the students, opportunities for the students to discuss strategy and choices, rationalization for wanting to improve decision-making, and justification for studying analytical techniques, such as optimization and prediction.

Selection of a Game

In order for a game to be worthwhile, the playing of the game must further the Student Learning Outcomes. Linking student learning outcomes to gameplay can be done easily. Game components should induce student engagement in decision-making. This decision-making during the game is the basis of future learning. Typical components are competition, limited information, a means of gauging progress, and well-defined victory conditions. For example, limited information can come in the form of shuffled deck of cards that is used to determine new conditions for each turn. This deck of cards is a source of information that is quite limited for the players. As the cards are revealed during the game, the students can track that information. As the amount of information from the cards increase (more data points), the students could simulate the randomness of the cards in various ways. Let's say the cards revealed the price of raw materials per turn. If the students were to graph the prices versus time in a scatter plot, the students could then calculate mean, and standard deviation, as well if the plot is skewed. Depending on the shape of this plot, the students could create random number generators (rng) from what they gathered from playing game. The students could use uniform, normal, or Poisson generators and see if their rng matches the cards' actual distribution of outcomes.

The number of positions, or players, is given by the published game; however, a position in the game can be replaced by a team of players. This change naturally increases the amount of time required to play the game because of the amount of discussion required to make decisions, but this discussion is sought for the learning opportunities. Therefore, games should be converted to teams if at all possible. Avalon Hill's Business Strategy is for four players. I've played the game several times where each player was converted to teams of four to six students. See the "A LISTING OF PUBLISHED GAMES" section for a description of this game, and the "EXAMPLE ONE—RUNNING A FACTORY" section for an example of this game being played in the classroom.

An important point to consider is the length of the game in turns and in hours. One game may typically take several hours to play with a minimal number of players. Increasing the number of positions and the number of individuals per position and the amount of times increases drastically. With everything considered, completion of the game is always secondary to learning. Ending a game at a certain time, say the end of the class period, may be acceptable to the players if the reason why the game is played is tied to the subject matter of the course. The point of the game must be given to the students explicitly.

Introducing the Game to the Students

Before play can begin, the students must become familiar with the victory conditions, rules of play, sequence of events, and options. Explaining the rules can take place outside of class, such as watching a video of a game being played, handouts, or a short lecture before playing. A rules summary for quick reference during a game helps tremendously in teaching how to play. The first turn can also be used to teach the sequence of events and options. Usually, the first turn takes much longer to play than sequential turns and cannot be used to estimate the number of turns that can be played in a given amount of time.

For the Business Strategy game board, I have created PowerPoint slides and projected that on a white board, then used dry erase markers to note positions on the board instead of actual game pieces. If cards are dealt to the teams, actual cards can be handed to the students. Or, as an alternative, a sheet with all cards explained can be handed to the students and then the students are told which ones are in their hands.

An alternative to all students playing a single game, the students can be divided into groups and each group will play one instance of the game. I have done this style of game play for the recreating the Battle of Hastings. I first created five or six game boards from poster board and created the playing pieces from foam board. Students were placed in groups of 4 or 5. Each group was split into the two armies at the battle. One advantage to this style of play in a classroom is that discussion after the game involves the students discussing the varying outcomes: Who won? What strategies were used? Any surprises?

Another option is to carry over the game into more than one class session. This allows team members to learn from their experiences and develop a new strategy before resuming the game. This is quite effective if the plan is to have a discussion about the game after playing. One class session is devoted to introducing the rules and playing one or two turns and then time between class sessions allows for team discussion. When class resumes, the first half of the class session is devoted to playing another couple of turns, and the second half is devoted to a general discussion, which could lead into future lecture topics.

Depending on the mechanics and rules of play, I suggest recording moves each turn by each team electronically. Turns can be submitted via the online Learning Management System, such as Blackboard, Canvas, or Moodle. Electronic submission of turns allows the turns to be recorded for later analysis. Recording the moves generates data that can be linked to outcomes by turn and the ultimate outcome of the game. Thus, all students in the classroom have a shared experience and empirical data.

Some games limit the amount of resources available to the players/teams on the first turn. This limitation on resources means that players also have limited options and the first turn's moves are rather simple, which means the first turn can proceed quickly. The rules can be taught in the first turn. If so, the team players need time to organize and learn the rules, which means the first turn can actually take longer to complete. More complex rules can be introduced as the game progresses.

In my experience, all team members playing the game are continually seeking more information and greater understanding of the rules and victory conditions. This information seeking behavior opens up discussion to cover topics such as sensemaking, decision-making, game theory, and communication theory, which can be combined with typical analytics pedagogy.

A Listing of Published Games

The games that I choose to play in the classroom are published games based on business and economic issues or historical battles. The actual topic covered is sometimes important, such as a game on the stock market, but usually not. If I want to use stock market data in one of my lectures, I can have the students play a game first and then download stock market data for later predictions. By playing the game, I know that all students have a fundamental understanding of the stock market before jumping into the data and building predictive models. As students' confidence in understanding topics, such as the stock market, increases so does the quantity and variety of questions asked about such topics during classroom discussions.

I have played the first three games in the list below with students multiple times. The last two in the list are games that I am considering. Unfortunately, some of these games are no longer published, but are available from second hand dealers and auction websites.

Business Strategy, 1973, Avalon Hill Game Company, Baltimore, Maryland

It is a four-player game on managing the raw resources and production of factories. Players bid for purchasing raw resources and selling finish goods. It has four levels of complexity: Family Game, Basic Business Game, Corporate Game, and Classroom Game.

Stock Market Game, 1970, Avalon Hill Game Company, Baltimore, Maryland

Players purchase Blue Chip, Speculative, Preferred, Bonds, and Warrants. The number of players is not limited. Prices of stock vary based on card play and players' purchases. It has two levels of play: Basic and Advanced. In the Advanced game, players can buy on margin, convert stock, and sell short.

Ancient and Medieval Wargaming, 2007, Neil Thomas, The History Press, Stroud, Gloucestershire

Players command armies and recreate historical battles on tabletop. The book is only the rules; playing pieces must be developed or purchased separately. Rules are given for building specific armies according to several lists, playing the game, and developing scenarios.

Executive Decision, 1971, 3M Company, St Paul, Minnesota

It is a two- to six-player game on managing the raw resources and production of factories. Players bid for purchasing raw resources and selling finished goods. Prices fluctuate based on demand of goods by the players. This game is similar to Avalon Hill's Business Strategy, but raw materials have different levels of quality and can handle more players.

Trailblazer, 1981, Metagaming, Austin, Texas

Players take control of space-faring corporations in attempting to make a profit by exploiting natural resources in other star-systems and selling them to consumers on Earth. This game is designed for two to four players and takes at least 4 h to play. The game requires a large board for the star map, counters for ships and planets, and dice for random number generation. The players use several spreadsheets to track production, exploration, sales, and maintenance. It is a very detailed game with lots of information.

Example 1: Running a Factory

The game I've played the most is Avalon Hills Business Strategy. In this game, the students manage a company that owns several factories turning raw materials into finished goods. The students must purchase raw material, build factories, and sell their products. Competition in the game centers on bidding for raw materials and setting prices for the finished products. The published game has a board, markers, cards, and spreadsheets.

A turn in the game represents 1 month of real time and has several steps. First, I would tell the students how many raw material units are for sale and the minimum price for purchase. The students, among the teams, would discuss privately a bidding price for these goods. Since the number of raw material units was set, the students quickly realize that raw materials are a precious commodity. Without raw materials, there will be no finished goods and factories will sit idle. The objective in this bidding phase is to set a bid high enough to capture raw material units, but not too high. Costly raw materials send finished goods prices soaring. The turn sequence also has phases for building new factories, upgrading factories, and selling finished goods. Factory purchase and upgrading is not competitive. Teams can build as many as they want. Selling finished goods is competitive and the teams must decide selling prices. Too high and no goods will be sold; too low and the company loses money.

To set up this game for the classroom, I recreated the board using Microsoft PowerPoint and then projected the board on a white screen in the classroom. I also built spreadsheets in Microsoft Excel based on the paper-based spreadsheets that came with the game. Then, I wrote a summary of the rules using Microsoft Word. Electronic copies of these were distributed to the students prior to the class session through the Learning Management System. I also recorded a video of a lecture describing the game.

At the class session, I projected the board on the screen and handed out the rules summary and the spreadsheet to the every student. I also divided the students randomly into four teams: Orange, Green, Blue, and Red. As we played the game, I would write on the white board important information, such as owned factories and the number of finished goods for sale. The students could see major information for each team/company. See Figure 6.1 for an image of the projected board. The spreadsheet was organized so that each column was a turn in the game. Students would complete the spreadsheet as the game progressed. Each team/company started with two factories, four raw materials, and four finished goods and $10,000. See Figure 6.2 for a copy of the spreadsheet. The goal of the game is to make more money than other teams.

Example 2: Warfare in the Age of William the Conqueror

Recreating the Battle of Hastings in the classroom has always generated a lot of enthusiasm and enjoyment for the students.

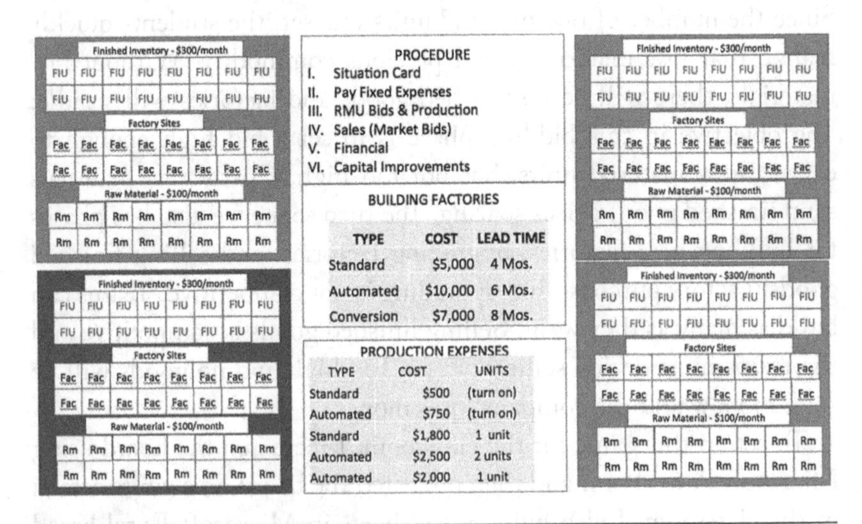

Figure 6.1 Recreation of the board for Avalon Hill's Business Strategy game for a classroom.

		Jan	Feb	Mar	Apr	May	Jun	Jul	Aug	Sep	Oct	Nov	Dec
II	**FIXED EXPENSES**												
a	Raw Material ($100/unit)												
b	Finished Goods ($300/unit)												
c	Turn on Std Fact. ($500)												
d	Turn on Auto Fact ($750)												
III	**RAW MATERIAL BIDS**												
a	Nbr of Units Bid												
b	Bid per Units												
c	Units Purchsed												
d	RM Total Cost												
III	**PRODUCTION COST**												
e	Std Fac Processing												
f	Auto Fac Processing												
g	Processing Costs												
IV	**SALES (MARKET BIDS)**												
a	Nbr of Units Offered												
b	Asking Price												
c	Nbr of Units Sold												
d	Total Income												
V	**FINANCIAL**												
a	Amount of Loan												
b	Interest Due												
c	Total Owed												
VI	**CAPITAL IMPROVEMENTS**												
a	Std Fac Units Ordered												
b	Auto Fac Units Ordered												
c	Conversions Due												

Figure 6.2 Recreation of the spreadsheet for Avalon Hill's Business Strategy game for team play in a classroom.

Instead of playing a single game with the board projected on a wall, I decided that students would play several games at the same time at individual tables. To do this, I asked students to separate into groups of four or five per table. The students at each table were split into two opposing factions, either Normans or Anglo-Danish. (Side note: The inhabitants of modern day England were more Anglo-Danish in 1066 than English.) Thus, several instances of the battle were being played in the classroom.

The Battle of Hastings is where William the Conqueror, from Normandy, won a decisive battle over Harold Godwinson in claiming the crown of England. The battle had Harold and his army of Anglo-Danish soldiers defending a hill and William's forces of knights, soldiers, and crossbowmen trying to take the hill. Historically, William almost lost the battle. It took him all day to wear down the defenders on the hill so that his knights could exploit a gap and destroy the Anglo-Danish army. Students refight this battle using rulers to measure movement and shooting, dice to resolve the effects of combat, and placement of army units to determine who controls the hill.

Each table of four to five students needed a game board, playing pieces, and sets of charts to recreate the battle. The game boards were maps of the battle site. I bought regular green poster boards and brown construction paper. I cut out football shapes from the construction paper and glued it on one of the long sides of the poster boards. This brown shape represented the hill. See Figure 6.3 for a depiction of the playing surface. Playing pieces represented units of soldiers, much like the playing pieces on a chessboard.

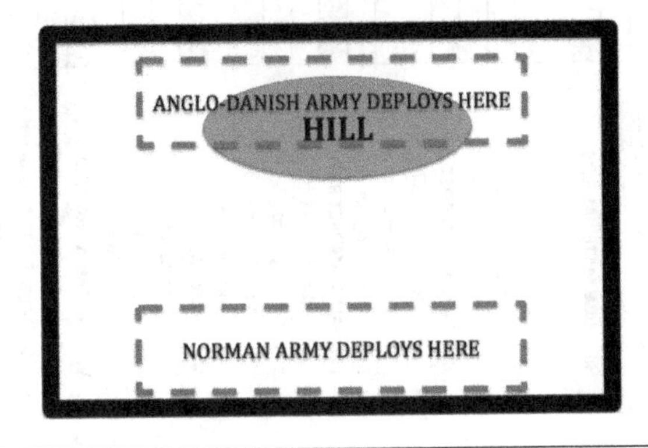

Figure 6.3 The map, or game board, for the Battle of Hastings.

To create the playing pieces I used Microsoft PowerPoint to design the units and printed out several copies. Each sheet had several units designed. See Figure 6.4 for the playing pieces. I bought foam board and cut it into several rectangles. Then, I glued a playing piece from the printed sheets onto each foam board rectangle. After all this construction, I had several Norman and Anglo-Danish armies. Each Norman army had nine playing pieces and each Anglo-Danish army had eight. The charts described how far individual units could move and how to determine combat outcomes. Thus, on each table were a poster board, rulers, dice, playing pieces, and charts sitting ready for the students.

As the students played the game, I moved from table to table to make sure everyone understood the rules and were making progress in the turns. Lots of questions came up during the games as students started to put together the rather complex rules into a strategy to either defend the hill or take the hill. The charts contained information concerning movement and combat capabilities. In comparison to the game on how to run a factory, as described above, the students had access to much more information, almost too much. Absorbing this information individually and as a team was a barrier for the students to develop tactics and strategies. The difference between having too little information and too much information in decision-making was discussed after playing.

Debriefing the students after this game, from an instructor's point of view, was very interesting. Several students combined special capabilities of a few the units into a coherent strategy. Some did not. Some strategies were more complicated than others. Students seemed

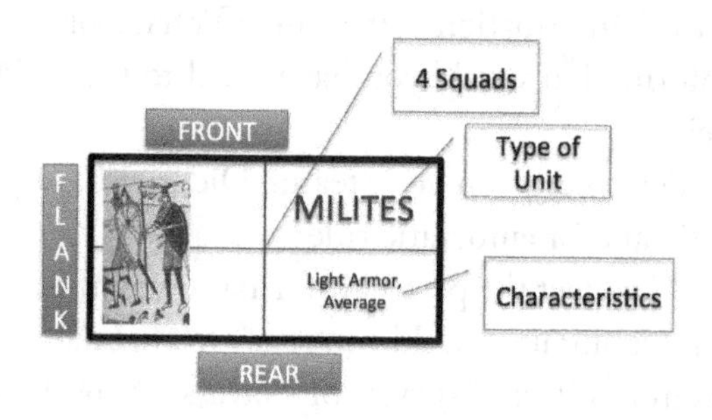

Figure 6.4 A unit of Norman soldiers, called Milites in the game, showing how to interpret the information on each playing piece.

to enjoy discussing different strategies and explaining how and why these strategies were developed. Also, the charts were discussed in how to better organize them and explain them, which led to discussions on data visualization. We also had discussions on information theory, communication theory, decision theory, and game theory.

Debriefing the Students after the Game

The most critical phase of playing games in the classroom is not the game, but the discussion immediately following the game. This discussion, or debrief, is where the actions and choices in the game are linked to future topics, lectures, and assignments. The discussion starts the questions that should be asked prior to the analytics work. Discussion topics after a game can include the decision-making process, team debates, the desire for information, critical decision points, expected outcomes and expected values, and how to improve strategies. Some of typical questions I ask after a game are:

1. How confident were you in the topic of the game before the playing and after playing?
2. How did your discussions compare to the Observe-Orient-Decide-Act (OODA) loop?
3. How difficult were decisions on your team and what made the decision harder or easier?
4. Did you have a strategy at the beginning? Or, did a strategy evolve over a few turns?
5. Did you calculate probabilities before making a decision? Why or why not?
6. Did you track information in the game? What tools did you use?
7. What information would you have liked to track? Why? For what use?
8. Did a leader emerge on your team? Did you have periods of democratic and/or autocratic rule?
9. If you could collect the perfect information on the game, what analytical techniques would you apply to that data?
10. If you could collect a survey of players of the game, what questions would you ask that could help you develop a winning strategy?

11. Did you learn everyone's names on your team? Was it important to learn everyone's names?
12. Did the outcome of the game (who won) surprise you?

The discussion of the game, understanding strategy, and choices made by the players is critical to setting up future lectures and topics. The discussion after game can create links to topics, such as association analysis, validity, and reliability, and hypothesis testing. For example, if the next planned topic in the course is probability distributions and hypothesis testing, I may generate some dummy data about the game or have empirical data from previous game sessions, then steer the discussion toward posing a hypothesis about some action in the game. The students would then use that data to test the hypothesis.

Homework Assignments Based on the Game

If a course uses stock market data, the students may benefit from playing a stock market game. The students could become familiar with buying and selling and risk. From that discussion future homework assignments could be on optimization, simulation, or predicting the winner in the middle of the game. Having the students describe the game turn phases and the decision making process can lead to interesting and insightful theories about how to win in the game. The shared experience of the game enables students of varying backgrounds to collaborate on writing research questions, which leads to posing hypotheses or building predictive models. As the discussion grows, students should brainstorm techniques that tie analytics techniques to specific game actions.

Students would then have several homework assignments: (1) collect data, (2) perform exploratory data analysis, (3) develop models, and (4) evaluate models. It is even possible to return to playing the game again at the end of the semester to "test" these new models. Students could present theories on the game in a poster session. Students could develop a survey about playing the game, cluster the results of the survey and choices made during the game, and then determine what can be learned from the survey data. Two other examples are provided below in a detailed manner.

Data on bidding in the Business Strategy game was collected along with the status of the companies, such as the number of raw materials, finished goods, and the number of factories. The students were asked to create a model that predicted a company's bid given the information about that company and the other companies. Most models produced were poor in accuracy due to the small amount of data and poor collection methods. Once the models were evaluated, I asked the students to explain why. Then, I asked the students to come up with ideas on how to improve the models through other data collection techniques, list of desired variables, and possibly new algorithms for producing new models.

Homework Example 1: Optimization

One example I have used from the Ancient and Medieval Wargames is the rules regarding army building. Historical armies differ by composition of troops. For example, Hannibal, circa 218 BCE, crossing the Alps had a few elephants. An army of Genghis Khan, circa 1200 CE, had mostly cavalry. Each of these historical armies has a few options, which give players the ability to tailor each army slightly. Once the historical army is chosen, a player then could design an army to suit her or his needs.

After playing the game, I pick one army and ask the students to build that army according to the rules using a specific goal of optimizing an army feature. A typical army in the game gives several options. The Parthian army can have between two and four Noble heavy cavalry, between four and six light cavalry, and up to two militia units. An army of Alexander the Great has many more options: two types of heavy infantry, two types of light infantry, heavy cavalry, and light cavalry. Each type in the Alexandrian army has different minimums and maximums.

Depending on a player's particular style in playing the game, a player could choose to play a particular kind of army. Designing the army then becomes an optimization problem. See Table 6.1 for a listing of army unit types with minimum and maximum amounts for an Alexandrian army.

Each unit has a different movement rate and combat capabilities in the game. These movement rates and combat capabilities could be counted, summed, or averaged across the whole army in order

Table 6.1 Minimum and Maximum Amounts of Units for One Particular Army in the Game

ALEXANDRIAN MACEDONIAN ARMY			
UNIT TYPE	HISTORICAL NAME	MINIMUM	MAXIMUM
Heavy infantry	Phalangites	3	5
Heavy infantry	Hypaspists	0	1
Light infantry	Agrianians	1	2
Light infantry	Cretans	0	2
Heavy cavalry	Companions	1	2
Light cavalry	Thessalians	0	1

to describe that army's capabilities. For instance, the Parthian army could be considered a very mobile army because of the high movement rates of the cavalry. The Alexandrian army could be considered an offensive army because of the amount of heavy infantry.

Optimization problems are traditionally presented as challenges to students to calculate the best mix of options to maximize or minimize some variable. One typical optimization problem for students is to maximize profit when presented with options in production of sellable goods. Another problem is to minimize the number of employees to hire given a work schedule that needs a minimum number of employees per shift. Optimization can be performed in spreadsheets, such as Microsoft Excel or Libre Office Calc. Many other statistical packages exist that can perform optimization. Likewise, students could design an army using optimization.

Through optimization, a player could try to achieve maximum rates of movement, the defensive features of the army, or missile fire capability. These optimization problems are not simple problems for students. Part of the difficulty in this optimization problem for students is that the data are strung out among many pages of documentation and there is no simple rule for what information to include and not include. The list of choices adds to the complexity of solving the optimization problem. Thus, students are not presented a classic optimization problem; instead, they are presented with their own strategy of army choice, which naturally grows into an optimization problem.

Homework Example 2: Simulation

Another example using the same game rules is creating a simulation using the Python programming language. Once an army is built using optimization I ask the students if it is possible to guess how the army would perform against prebuilt typical opponent armies. This leads to a discussion on how to determine the strength of an army without actually playing a full game. Determining army strength should be based on a literature review of available material. Once a method has been determined, I have the students program the method in Python and run the method many times. This introduces the students to random number generation, simulation, Monte Carlo methods, and the like.

The particular method I used in teaching simulation is where unit combat strength is summed and then a simple ratio of the two armies is created. The probability of one army winning in a particular matchup is then computed as follows:

$$P(\text{Winning}) = (\text{army combat strength}) / (\text{army combat strength} + \text{opponent combat strength}).$$

I also add in a few modifiers to combat strength depending on the match up. I add these modifiers mostly for the purpose of showing my features of the Python programming language. Depending on the students' programming capabilities, students can write the program from scratch or I give them a Python routine and ask them to modify it. One modification to the routine the students must perform is to run a combat at least 1,000 times and write the results to a file. See Appendix 6A for some base python code that could be handed to the students to debug and modify if the students' skills with python are limited.

The idea behind both the optimization and the simulation homework problems is to pick a winning army for an imaginary tournament. The students are forced to make choices among armies for the tournament. In making this choice, the students analyze the strengths of the armies based solely on numeric strength at first. Then the students can take some theories (gained from qualitative research) into account in our analysis and perform a simulation. Then, the students are able to practice with the chosen army in their own time.

Outcomes from Playing Games in the Classroom

Other benefits to the students are possible with playing games in the classroom. Students who have been in courses that use the case method of teaching have mentioned that playing games gives them the ability to ask other students about choices made. In a case study, the students do not have access to the actors in the case. However, the students do have direct access to fellow players in the game. Playing games also gives the students one more avenue of contact with fellow students. Thus, students have the opportunity to expand their individual networks. Students have the ability to show different skill sets within the classroom, which could lead to building self-confidence. Playing games also creates an open cooperative environment, which leads to increased student engagement. The fun of learning comes out. As the students participate in shared activities, students perceive the contributions made by each other, and bonds between students are formed.

After playing the games, I asked students to complete a survey. Each game had its own survey, so that some questions were unique to each game and some questions were universal to all games and appeared on all the surveys. I also conducted a survey at the end of each semester asking the students to rate the different activities during the semester. In any given course, I usually give plenty of lectures, homework assignments, forum questions and discussions, and the games. At the end of the semester, I asked the students which activities were the most helpful and least helpful in understanding the theories and application of the theories presented in class. I also asked which activities reinforced the theories. The following tables show how the students responded to those questions.

Of those who said Games are most helpful, Homework tended to be the least helpful. Of those who said Homework was the most helpful, which was only one student, the games were the least helpful. Of those who said the lectures were the most, students tended to be split among the games and the homework being the least helpful. Of those who said the lectures were the most, students tended to indicate that the forums were the second most helpful. (This is not shown on the chart.) See Table 6.2 for a comparison of counts of students and their opinions on the most and least helpful teaching techniques.

Of the students who said that Games were the most helpful, most students said the games reinforced the topics – not that surprising. About a third who said games reinforced the topics also said that the reinforcement came eventually, not immediately. Of the students who said that Lectures were the most helpful, tended to agree that the

Table 6.2 Comparison of Students' Opinions on MOST Helpful to LEAST Helpful Teaching Technique

		RATED LEAST HELPFUL					
		ALL	FORUMS	GAMES	HOMEWORK	LECTURES	TOTALS
RATED MOST HELPFUL	ALL	0	0	0	1	0	1
	FORUMS	0	0	4	5	0	9
	GAMES	0	4	1	10	0	15
	HOMEWORK	0	0	1	0	0	1
	LECTURES	1	1	5	4	0	11
	TOTALS	1	5	11	20	0	**TOTALS**

games did reinforce the topics. About 10% of the students who enjoyed the games thought the games were unrelated. See Table 6.3 for a comparison of counts of students and their opinions on the most helpful teaching techniques versus which techniques helped reinforce the course topics. In Table 6.3, "Eventually" means that the students saw the connection between playing the games and the course topics after a bit of time. "Other Ideas" means that the students saw the games as reinforcing ideas other than those discussed in class. "Reinforced" means that the student saw the games and the connection to the course topics immediately. "Unrelated" means that the student did not see a connection between playing the games and the course topics.

Of all the students who thought the games were least helpful, only one thought the games were unrelated to the topics (see Table 6.4). Far more students caught onto the connection between the games and the topics than those who did not. And among the students who saw the connections, more saw the connections quickly.

Table 6.3 Comparison of Students' Opinions on MOST Helpful Teaching Technique Against How the Games Reinforced the Course Topics

		GAMES REINFORCED THE TOPICS?				
		EVENTUALLY	OTHER IDEAS	REINFORCED	UNRELATED	TOTALS
RATED MOST HELPFUL	ALL	0	0	1	0	1
	FORUMS	4	2	5	0	11
	GAMES	6	1	10	2	19
	HOMEWORK	0	0	1	0	1
	LECTURES	7	1	6	1	15
	TOTALS	17	4	23	3	**TOTALS**

Table 6.4 Comparison of Students' Opinions on LEAST Helpful Teaching Technique Against How the Games Reinforced the Course Topics

		GAMES REINFORCED THE TOPICS?				
		EVENTUALLY	OTHER IDEAS	REINFORCED	UNRELATED	TOTALS
RATED LEAST HELPFUL	ALL	1	0	0	0	1
	FORUMS	3	0	2	0	5
	GAMES	3	2	4	1	10
	HOMEWORK	7	1	11	1	20
	LECTURES	0	0	0	0	0
	TOTALS	14	3	17	2	**TOTALS**

Conclusion

One goal I have in teaching analytics courses is to engage the students in dialogue about the application of analytical techniques to a given problem set. One obstacle to getting all students to participate in the conversation is having a problem set that appeals to all students. A typical course in analytics can have students from finance, biology, computer science, and many other fields. Therefore, a shared experience in decision-making, especially if formed into teams, can overcome this diversity. Playing games is one solution to the problem of creating a data set that generates interest among students with diverse backgrounds.

Appendix 6A: Base Python Code for the Simulation

```
def
# Python routine for simulation on Neil Thomas'
Ancient Warfare Rules
#

import random


# This routines compares two armies using the tactics
table from Imperator game (World Wide Wargamer, 1990)
# The first army could be the player's army
# while the second army is the enemy

# First Army Units - all units default to a hardcoded
1, these should be modified to a particular army
```

```
A1Eleph = 1
A1Char = 1
A1HvInf = 1
A1Wrbnd = 1
A1Aux = 1
A1HvCav = 1
A1HvChar = 1
A1Archr = 1
A1LtCav = 1
A1LtInf = 1
A1Art = 1

# Second Army Units - all units default to a hardcoded
1, these should be modified to a particular army
A2Eleph = 1
A2Char = 1
A2HvInf = 1
A2Wrbnd = 1
A2Aux = 1
A2HvCav = 1
A2HvChar = 1
A2Archr = 1
A2LtCav = 1
A2LtInf = 1
A2Art = 1

# First Army Unit Classes
A1HCLC = A1HvCav + A1LtCav
A1Infantry = A1HvInf + A1Wrbnd + A1Aux + A1Archr +
A1LtInf
A1AllMntd = A1Eleph + A1Char + A1HvCav + A1HvChar +
A1LtCav
A1AllUnits = A1Eleph + A1Char + A1HvInf + A1Wrbnd +
A1Aux + A1HvCav + A1Archr + A1LtCav + A1LtInf + A1Art

# Second Army Unit Classes
A2HCLC = A2HvCav + A2LtCav
A2Infantry = A2HvInf + A2Wrbnd + A2Aux + A2Archr +
A2LtInf
A2AllMntd = A2Eleph + A2Char + A2HvCav + A2HvChar +
A2LtCav
A2AllUnits = A2Eleph + A2Char + A2HvInf + A2Wrbnd +
A2Aux + A2HvCav + A2Archr + A2LtCav + A2LtInf + A2Art

# First Army's initial probabilities per tactic
```

```
# Note that these really aren't probabilities ranging
from 0 to 1
# It's more like odds, the higher the number, the more
likely the outcome
A1ChargeProb = 1
A1RefuseProb = 1
A1EnvelopeProb = 1
A1StandProb = 1

# Second Army's initial probabilities per tactic
A2ChargeProb = 1
A2RefuseProb = 1
A2EnvelopeProb = 1
A2StandProb = 1

# Next Modify the probabilities by comparing army
classes
# Modify probabilities for the first army
if A1HCLC > A2HCLC: A1EnvelopeProb = A1EnvelopeProb + 2
if A1AllMntd > A2AllMntd: A1ChargeProb = A1ChargeProb + 1
if A1Infantry > A2Infantry:
      A1RefuseProb = A1RefuseProb + 2
      A1StandProb = A1StandProb + 1
if A1Infantry < A2Infantry:
      A1RefuseProb = A1RefuseProb - 1
      A1EnvelopeProb = A1EnvelopeProb - 1
A1TotalOptions = A1ChargeProb + A1RefuseProb +
A1EnvelopeProb + A1StandProb

# Modify probabilities for the second army
if A2HCLC > A1HCLC: A2EnvelopeProb = A2EnvelopeProb + 2
if A2AllMntd > A1AllMntd: A2ChargeProb = A2ChargeProb + 1
if A2Infantry > A1Infantry:
      A2RefuseProb = A2RefuseProb + 2
      A2StandProb = A2StandProb + 1
if A2Infantry < A1Infantry:
      A2RefuseProb = A2RefuseProb - 1
      A2EnvelopeProb = A2EnvelopeProb - 1
A2TotalOptions = A2ChargeProb + A2RefuseProb +
A2EnvelopeProb + A2StandProb

# Next choose the tactics for both armies
# first we initialize some variables to 0, indicating
no tactic has been chosen yet
A1Charge, A1Refuse, A1Envelope, A1Stand = 0, 0, 0, 0
```

```
A2Charge, A2Refuse, A2Envelope, A2Stand = 0, 0, 0, 0
# set our seed for the random numbers
# random.seed(162892) # can set the seed if you want
# next generate a random number and compare that to
some ratios
A1TacticsRG = random.randint(1, A1TotalOptions)
A2TacticsRG = random.randint(1, A2TotalOptions)

# next determine the tactic for each army
A1Tactic = 0
if A1TacticsRG <= A1ChargeProb:
      A1Tactic = 1 # charging
      print("Army 1 is charging.")
elif A1TacticsRG <= (A1ChargeProb + A1RefuseProb):
      A1Tactic = 2 # refuse
      print("Army 1 is refusing.")
elif A1TacticsRG <= (A1ChargeProb + A1RefuseProb +
A1EnvelopeProb):
      A1Tactic = 3 # envelope
      print("Army 1 is enveloping.")
else:
      A1Tactic = 4 # stand
      print("Army 1 is standing.")
A2Tactic = 0
if A2TacticsRG <= A2ChargeProb:
      A2Tactic = 1 # charging
      print("Army 2 is charging.")
elif A2TacticsRG <= (A2ChargeProb + A2RefuseProb):
      A2Tactic = 2 # refuse
      print("Army 2 is refusing.")
elif A2TacticsRG <= (A2ChargeProb + A2RefuseProb +
A2EnvelopeProb):
      A2Tactic = 3 # envelope
      print("Army 2 is enveloping.")
else:
      A2Tactic = 4 # stand
      print("Army 2 is standing.")

# What is the outcome of comparing the two tactics?
# in other words, which troops of which army is
doubled or tripled?
# first army charges
if (A1Tactic == 1 and A2Tactic == 1):
      A1Fight = A1AllUnits + A1AllMntd
      A2Fight = A2AllUnits + A2AllMntd
```

```
if (A1Tactic == 1 and A2Tactic == 2):
     A1Fight = A1AllUnits
     A2Fight = A2AllUnits + A2AllUnits
if (A1Tactic == 1 and A2Tactic == 3):
     A1Fight = A1AllUnits + A1HCLC + A1HCLC
     A2Fight = A2AllUnits + A2HCLC + A2HCLC
if (A1Tactic == 1 and A2Tactic == 4):
     A1Fight = A1AllUnits + A1AllUnits
     A2Fight = A2AllUnits
# first army refuses
if (A1Tactic == 2 and A2Tactic == 1):
     A1Fight = A1AllUnits + A1AllUnits
     A2Fight = A2AllUnits
if (A1Tactic == 2 and A2Tactic == 2):
     A1Fight = A1AllUnits
     A2Fight = A2AllUnits
if (A1Tactic == 2 and A2Tactic == 3):
     A1Fight = A1AllUnits + A1AllUnits
     A2Fight = A2AllUnits
if (A1Tactic == 2 and A2Tactic == 4):
     A1Fight = A1AllUnits + A1Infantry
     A2Fight = A2AllUnits + A2Infantry
# first army envelopes
if (A1Tactic == 3 and A2Tactic == 1):
     A1Fight = A1AllUnits + A1HCLC + A1HCLC
     A2Fight = A2AllUnits + A2HCLC + A2HCLC
if (A1Tactic == 3 and A2Tactic == 2):
     A1Fight = A1AllUnits
     A2Fight = A2AllUnits + A2AllUnits
if (A1Tactic == 3 and A2Tactic == 3):
     A1Fight = A1AllUnits + A1HCLC + A1HCLC
     A2Fight = A2AllUnits + A2HCLC + A2HCLC
if (A1Tactic == 3 and A2Tactic == 4):
     A1Fight = A1AllUnits + A1Infantry
     A2Fight = A2AllUnits + A2Infantry
# first army stands
if (A1Tactic == 4 and A2Tactic == 1):
     A1Fight = A1AllUnits
     A2Fight = A2AllUnits + A2AllMntd
if (A1Tactic == 4 and A2Tactic == 2):
     A1Fight = A1AllUnits + A1Infantry
     A2Fight = A2AllUnits + A2Infantry
if (A1Tactic == 4 and A2Tactic == 3):
     A1Fight = A1AllUnits + A1Infantry
     A2Fight = A2AllUnits + A2Infantry
```

```
if (A1Tactic == 4 and A2Tactic == 4):
      A1Fight = 0
      A2Fight = 0

# Next calculate the odds of winning
if ((A1Fight + A2Fight) == 0):
      print("uh oh...No Battle Today!")
else:
      WinningOdds = (A1Fight + 0.0) / (A1Fight +
A2Fight)
      BattleOutcome = random.random()
      if BattleOutcome <= WinningOdds:
            print("You won!")
      else: print("You lost!")
      print("WinningOdds is ",WinningOdds)
      print("BattleOutcome is ",BattleOutcome)

#print("A1Fight is ",A1Fight)
#print("A1Tactic is ",A1Tactic)

#print("A2Fight is ",A2Fight)
#print("A2Tactic is ",A2Tactic)
```

7

Student Competitions

Extending Student Experience Outside of the Classroom

YELENA BYTENSKAYA, KATHERINE LEAMING GOLDBERG, AND ELENA GORTCHEVA

University of Maryland University College

Contents

Introduction

Experiential learning has been recognized as a valuable tool for cementing student understanding and allowing them to experience near to real-world conditions in applying concepts learned in the classroom. This is especially important in light of employer expectations of skill levels for recent college graduates. The 2018 NACE Job Outlook Survey indicates a bit of disconnect between student perception of readiness and employer experience, uncovering some employer perception of lack of readiness pertaining to skills necessary for workplace success (Bauer-Wolf, 2018). A similar finding was reported in research conducted by the Association of American Colleges and Universities in 2015. Bauer-Wolf (2018) quotes the AAC&U report as stating that areas in which employers do not feel graduates are sufficiently well prepared is "particularly the case for applying knowledge and real-world skills in real-world settings, critical thinking skills, and written and oral communications skills" (para. 20).

Studies have found that experiential learning activities in the form of competitions might help to bridge this gap. For instance, a study of students participating in a Google Online Marketing Challenge found that learning outcomes included interpersonal life skills, digital technical skills, intrapersonal life skills, and adaptive applied skills (Croes & Visser, 2015). A host of such competitions providing experiential learning are available to college and university students studying analytics (see Resources at the end of this chapter). In this chapter, we will discuss lessons learned by educators at the University of Maryland University College's (UMUC) Master of Data Analytics program with guiding student participation in the IBM Watson Analytics Global Competition.

Dr. Elena Gortcheva, Program Chair and Professor of Data Analytics at UMUC, learned of the IBM Watson competition for students at a conference she attended in 2016. The goal of this competition was to encourage students to seek innovative solutions to global environmental issues using big data and the IBM Watson Analytics tool. Since UMUC already had a license for IBM Watson and the product was integrated into existing curriculum, this competition seemed to be a good fit. In deciding to get students enrolled in UMUC's Master of Science in Data Analytics involved with this opportunity, the department set an initial objective of focusing on student participation and completion of the required projects rather than on winning. Each team consisted of three students using public databases. Students exceeded departmental expectations, with one team placing second in the contest in the inaugural year, and two teams placing first and third in the second year (Dempsey, 2017).

UMUC took two different approaches in integrating this experience into student learning. In the first year (2017), Dr. Gortcheva enlisted Yelena Bytenskaya, the teaching assistant for the program, and Dr. Knode, professor of data analytics, in determining which students would be appropriate to approach. They selected students who were scheduled to be in the same section of the capstone course together and who had previously demonstrated high proficiency using Watson Analytics. In 2018, an open announcement was made to students inviting them to join the competition; in addition, faculty also recommended students they felt particularly suited for the program. In both years, the students had to choose their own topics based on a global environmental problem that they wanted to solve.

In both years, student teams were advised by Dr. Gortcheva, Ms. Bytenskaya, and Dr. Knode. Due to the terms of the competition, advisors were only able to help with technical difficulties of Watson Analytics and could not provide students with any feedback on their presentations; essentially their roles were those of mentors.

After the teams were formed, students were provided with the requirements for the competition, which were a video presentation, written paper, and access for the judges to Watson Analytics. The students self-appointed roles and autonomously managed themselves, using a website called Slack.com to manage the project and their communication with each other. They were required to find a public data set to use for their analysis and load it into Watson Analytics. Some students needed extra help learning the details of Watson Analytics that they had not previously used in their coursework.

Their advisors checked in with them regularly to check on progress and provide encouragement as well as reviewing the final products of the teams before submission to ensure that all of the requirements were completed (incomplete submissions would be automatically disqualified). One of the teams struggled with its video because the students wanted to show animations and live demonstrations of Watson Analytics, but this proved to be too difficult to record. The students did revise their projects because they often needed to simplify and condense their presentations. When creating verbal presentations, one of the advisors recommends that "less is more." Even with the revisions, the students still had enough time to submit their entries before the deadline.

Submissions were judged by a panel of judges using a rubric that the students were given in advance of the competition. The judging period took 1 month. In 2018, there was a technical issue with Watson Analytics, and an extension was provided to allow students enough time to work. After being judged, the students were notified via email that they were in the list of the top ten. Once UMUC knew that its students were in the top ten, university staff worked to make travel arrangements for the students. The final results were revealed in person in Langkawi, Malaysia (2017) and Shanghai, China (2018).

Some of the benefits of having the students participate in a competition are that students are able to measure their abilities against students at other universities with an actual project. In addition to the quantitative skills the students gained, they also developed their communication skills because they needed to express their ideas efficiently and clearly. The students also improved their ability to create video recordings and tell a story. This ability to create compelling narratives proved crucial to the high rankings the UMUC student teams achieved. The 2018 first-place team leveraged a personal experience of one of its members to bring immediacy to the team's analysis of the impact of air pollution in sub-Saharan Africa (Dempsey, 2018). The winning team is displayed in Figure 7.1. This ability to translate big data analysis to a call for action addresses the types of needs expressed by employers.

There are some risks associated with asking students to participate in competition. To begin with, if the students are spending too much time working on their competition, their coursework might suffer. Another possible risk could be that if the students are not able to complete the project, they might lose confidence in their abilities. As with any competition involving possible student travel, it is necessary to find funding to get finalist teams to the conferences at which the top winners are announced. Finally, these types of competitions are governed by outside

Figure 7.1 Photograph of University of Maryland University College teams.

parties and so faculty cannot assume that the format and structure of a competition will be continued into the future. The IBM Watson Global Competition represents an excellent example of this last risk as IBM is no longer marketing the Watson product and instead has converted to IBM Cognos Analytics ("IBM Watson Analytics", n.d.)

The UMUC advisors feel that their students' abilities improved as a result of the competition and so they are planning to continue asking students to join similar competitions in the future. For anyone interested in engaging their students in similar experiential learning to supplement classroom learning, the following steps are suggested to identify competitions appropriate to their programs:

1. Review Kaggle.com to determine if there are any appropriate competitions that would mesh with your existing curriculum.
2. Monitor www.kdnuggets.com/competitions/ for any appropriate challenges for your students.
3. Research any competitions offered by software packages students might already be using such as SAP, Tableau, SAS, and RapidMiner.

References

Bauer-Wolf, J. (2018, February 23). Overconfident students, dubious employers. *Inside HigherEd.* Retrieved from https://www.insidehighered.com/news/2018/02/23/study-students-believe-they-are-prepared-workplace-employers-disagree

Croes, J. V., & Visser, M. M. (Fall 2015). From tech skills to life skills: Google online marketing challenge and experiential learning. *Journal of Information Systems Education, 26*(4), 305–316.

Dempsey, M. (2017). UMUC team wins Watson analytics competition. Retrieved from https://globalmedia.umuc.edu/2017/07/11/umuc-team-wins-watson-analytics-competition/

Dempsey, M. (2018, June 18). UMUC students use big data to tell a personal story in winning IBM Watson Analytics Global Competition in Shanghai. *University of Maryland University Center Global Media Center.* Retrieved from https://globalmedia.umuc.edu/2018/06/18/umuc-students-use-big-data-to-tell-a-personal-story-in-winning-ibm-watson-analytics-global-competition-in-shanghai/

"IBM Watson Analytics" (n.d.). Retrieved from https://www.ibm.com/watson-analytics

Resources

Kaggle is an excellent way to find varieties of competitions as well as sample data sets. https://www.kaggle.com/

SAP hosts "hackathons" for its participants of the university alliances. https://events.sap.com/university-alliances-hackathons/en/home

RapidMiner software has competitions for its user base. https://community.rapidminer.com/t5/Data-Science-Competitions/Welcome-to-RapidMiner-Data-Science-Competitions/td-p/41414

This blog describes the submissions in the 2017 visualization contest sponsored by Tableau. https://www.tableau.com/about/blog/2017/12/and-winner-student-viz-contest-78794

KDNuggets has a dedicated web page for example datasets that students might want to use. https://www.kdnuggets.com/datasets/index.html

KDNuggets also has a webpage for existing data competitions for students and professionals. https://www.kdnuggets.com/competitions/index.html

SAS has a student competition using their software and can be found on this webpage. https://www.sas.com/en_us/events/sas-global-forum/program/awards-academic-programs.html#student-symposium

INFORMS (The Institute for Operations Research and the Management Sciences) has a competition for students using the same business problem, data sets, and software. https://www.informs.org/Recognizing-Excellence/INFORMS-Prizes/INFORMS-O.R.-Analytics-Student-Team-Competition

Adobe has hosted competitions for over 12 years for students using Adobe products with access to data from real-world organizations. http://adobeanalyticschallenge.com/

Teradata hosts a competition to solve a problem for a nonprofit organization. https://www.teradatauniversitynetwork.com/Community/Student-Competitions/2017/2017-Data-Challenge-Finalists

SECTION III

Program Design Tactics

8

Competencies for the Design, Implementation, and Adoption of the Analytics Process

EDUARDO RODRIGUEZ
University of Wisconsin-Stevens Point

JOHN S. EDWARDS
Aston Business School

GERMÁN A. RAMÍREZ
GRG Education LLC

Contents

Introduction

This chapter aims to review ways to develop competencies to implement the analytics process in organizations. There are two main streams to analyze the general problem of competencies improvement associated with the capacity to adopt analytics knowledge. On the one hand, organizations need to develop a continuous and ongoing learning process to use the new analytics knowledge in the most efficient and effective way. On the other hand, organizations need to gain competencies to contribute to the development of analytics knowledge and implementation of analytics solutions.

In this chapter, the concept of analytics refers to the practice and theory of what is presented in the literature variously as data analytics and business analytics. Additionally, the emphasis is on how to prepare analytics students to convert data into actions and solutions implementation. The main idea is to develop the skills to move from Know-What (problems and possible methods to solve them) and Know-Why (identification of possible sources of variation of results) to the Know-How to perform the analytics process (the value and impact of possible implementation of solutions) (Garud, 1997). Regarding Know-How, Charan (2007) pointed out eight skills that performers need to possess. These skills interpreted in the setting of the analytics process can be described as follows: first, understanding where the business is; second, keeping the business on the offensive (predictive analytics); third, creating the required analytics knowledge and sharing it; fourth, reviewing people's performance; fifth, keeping the analytics team competent; sixth, defining goals; seventh, aligning resources and goals; and finally, anticipating changes in the market (predictive and prescriptive analytics).

There is a lot of work to perform from basic technical skills to the adoption of the analytics process. The literature shows that competencies in algorithms use and technology literacy are essential for the analytics process development. Several surveys and studies have been conducted to understand what organizations need from analytics professionals. Some of those studies indicate aspects to review in the required talent in the present and future workplace. Price Waterhouse Coopers and the Business-Higher Education Forum (2017) indicate the need of analytics skills not only in the heads of professionals doing technical work but also in managers in different areas in organizations. The study differentiates between analytics-enabled jobs and data science jobs. The skills are classified as basic, intermediate, and advanced. In the category of analytics-enabled jobs, the most important skill is domain knowledge. Conversely, on the data science side, the requirements are about domain knowledge, machine learning, visualization, data governance, data management, and analytics approaches. In addition, the Asia-Pacific Economic Cooperation (2017) reviewed the shortage of analytics professionals and presented competencies in groups of: operational analytics, data visualization and presentation, data management and governance, domain knowledge and application, statistical techniques, computing, data analytics methods and algorithms, research methods, data science engineering principles, and what they called 21st-century skills.

However, the skills identified by the above surveys appear to fall more in the Know-What and Know-Why of the analytics process. In this chapter, the approach is based on what type of work is required to perform from a problem definition to the review of results and implementation of solutions (see Figure 8.1). The skills suggested here fulfill the need of performing and adopting the described tasks in the analytics process and workflow design and implementation.

Thus, keeping in mind Charan's skill set for Know-How, we propose to review not only competencies based on mastering technical aspects but also those related to the use of the available intelligences (note that we are using this term in the plural, as will be explained below) and the way to implement solutions. Regarding the development of the analytics skill set, we consider the conjunction of various intelligences, knowledge, and adoption of the analytics process. The starting point is to consider the development of intelligences as a process of adaptation of organizations to the evolution and revolutions of the conditions of competition, production, and operation. The main concept is that organizations are using data analytics as a way to be smarter and to create knowledge to mitigate the risks (develop competitiveness) associated with the ongoing changes in the business environment. Piaget noted: "Intelligence is an adaptation… To say that intelligence is a particular instance of biological adaptation is thus to suppose that it is essentially an organization and that its function is to structure the universe just as the organism structures

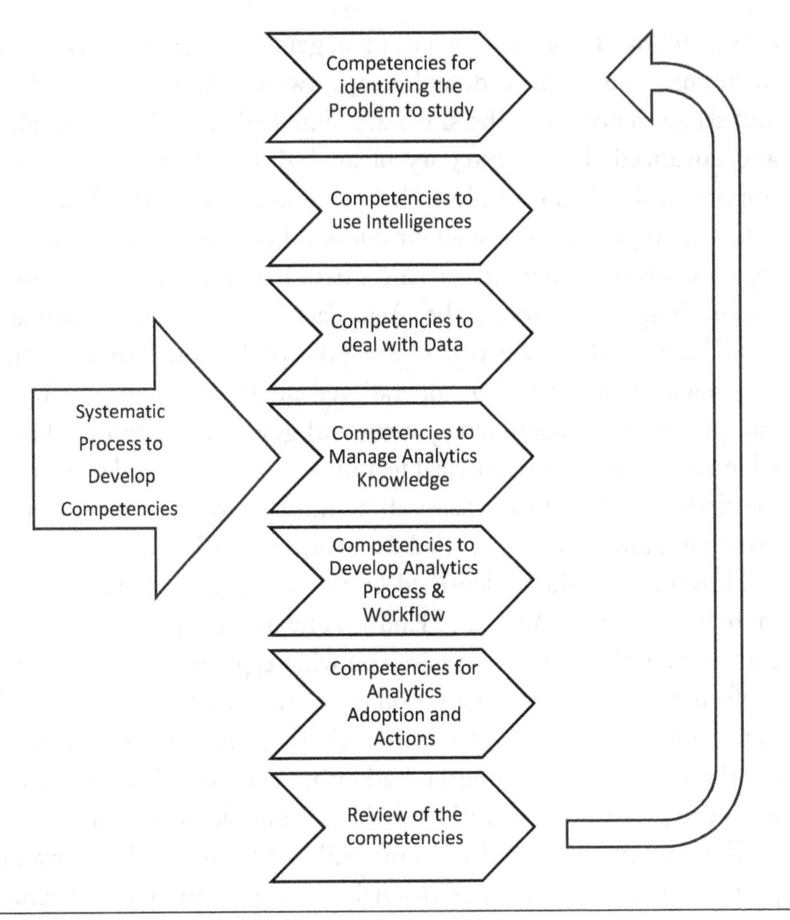

Figure 8.1 Development of skills-competencies in analytics.

its immediate environment" (1963, pp. 3–4). The adaptation of organizations to the business environment includes the appropriate use of data as a resource; as Henke et al. (2016) pointed out, "The value of data depends on its ultimate use, and ecosystems are evolving to help companies capture that value." There are use cases and sources of value, data, and modeling to generate analytics workflow. This analytics workflow provides the required changes of the organizations to new business conditions and the use of data analytics for adaptation to better performance.

Based on the concept of adaptation as the core aspect to review in this chapter, we have selected four concepts of intelligence as the pillars to support adequate search and use of the analytics knowledge that is created from data. These concepts are Sentient Intelligence,

Collective Intelligence, Strategic Intelligence, and Artificial Intelligence. The concepts of intelligences are considered the fuel of the analytics process/workflow and competencies building process. Competencies should be developed to generate appropriate statistical, computational, and business actions for analytics adoption in organizations.

In summary, the presentation of the general view of the competencies includes not only the outcomes of the surveys presented above but also the possibility to include new and deeper understanding of what the analytics professional requires to achieve the adoption of the analytics process in organizations. The subsequent sections are divided as shown in Figure 8.1.

Competencies for Identifying the Problem to Study

The competencies under this heading are as follows: define the problem, delimit it, identify the scope, define/formulate questions to answer and determine the variables to work with.

Problems in organizations need a complete study from the roots, scope, shape, and constraints in order to be considered appropriate to pursue with the analytics process. In several cases, the review of problems is on some of the symptoms only, and that practice can drive to limited answers and half-truths. There are areas in management that are guiding organizations to be clearer in the definition of problems such as Lean analysis, Six Sigma, Information Systems, and Operations Research/Management Science approaches. However, in analytics, there is a need to spend the required time to define the problem, to study and review/revisit it as a natural piece of a scientific study. People working on data analytics define the problem and identify the intelligences to use, as well as the methods to create analytics knowledge that provides meaning and actions for the problem resolution.

A clear definition of the problem provides the possibility to clarify the steps to follow in a systematic way to obtain insights, answers and guidance to move ahead in search of a clearer understanding of problems and solutions. The problem resolution in analytics is a chain of partial solutions that works in a similar way to the advance of science. To solve the problem of better performance in a particular key indicator may be a sequence of improvements in data management, data analytics, people's development, understanding of the impact, and value generation. The most important aspect is how the steps are built based on the previous ones in order to obtain better solutions in each of the advances or iterations.

The first step to define the skills required comes from answering the question "What are the problems that we would like to understand, to solve, to delimit, etc.?" An analogy is the problem of trying to reach an underwater treasure. Depending if the treasure lies at 1 m, 100 m or 1,000 m under water, the skills will be different, and the equipment to reach the treasure will be different as well. In analytics, the conditions to solve problems in marketing, risk management or other areas can require several levels of aspects to develop and to introduce in search of solutions.

Competencies to Use Intelligences

These competencies are the capacity to select and use to their full effect the intelligences at different levels of the analytics process.

In the introduction, we mentioned the importance of working with different intelligences in the development of the analytics process. In this section, we delve deeper into what we refer to as the intelligences that influence data analytics work. There are four intelligences to consider: sentient intelligence, collective intelligence, strategic intelligence, and artificial intelligence (Rodriguez, Edwards, & Facundo, 2011).

Sentient intelligence is a concept raised by the Spanish philosopher Xavier Zubiri between 1980 and 1983 in his trilogy *Intelligence and Reality*, comprising *Sentient Intelligence* (1980), *Intelligence and Logos* (1982) and *Intelligence and Reason* (1983). What Zubiri aims to address is the effect of the human environment and the way we as humans react in order to solve problems through the sociocultural relationships that exist in the environment and feelings and perception and apprehension. Apprehension for Zubiri means "I sense color and understand what this color is, too. In this case, the two aspects are distinguished not as types, but as distinct modes of apprehension" (Zubiri, 1999, p. 11).

It is important not to confuse sentient intelligence with emotional intelligence, the latter concerning the human capacity to feel, understand, monitor and modify *emotional states* in oneself and others, targeting a balance that allows to manage emotions intelligently. Proper identification of feelings involves a good understanding, but to know how to express and relate them to what makes them be, requires a more rational effort; that is, sentient intelligence.

For example, the appropriate relationship between sentient intelligence and visual analytics knowledge is part of the development of skills of analytics professionals. Pfaff (2016, p. 36) indicates, "Since our emotions shape our expectations of what might be there, the connections between sensation and emotion are clearer than ever."

Collective intelligence in analytics work represents that the search for solutions to problems is based on multidisciplinary work and the capacity of creating knowledge using different minds and perspectives on a problem's solution. A group solution to an analytics problem is not the sum of individual solutions, but the integration of individual solutions into a whole. Collective intelligence focuses on the resolution of problems through the use of collective knowledge. In the context of analytics we are not only interested in the source of the knowledge but also and mainly in the way that we create "ensembles" where the experts are learners that can grow knowledge faster, but the members (base learners), whether they are algorithms or people, provide additional learning capacities to solve the problem.

Strategic intelligence. A business organization is in search of means to achieve goals. The means to achieve goals are embedded in the strategy, and the strategy is based on analytics work to support the design and implementation of the strategy itself. Strategy and Strategic Intelligence is defined as (Service, 2006, p. 61) "a journey of planning, implementing, evaluating and adjusting while paying attention and focusing on the right things... (SQ_Strategic Intelligence) SQ_is the ability to interpret cues and develop appropriate strategies for addressing the future impact of these cues." It requires a strategic intelligence system (Montgomery & Weinberg, 1981) that provides information and knowledge for supporting and guiding the strategy formulation and implementation of organizations.

Artificial intelligence. Human–machine interactions are the norm for analytics development; humans create the means for machines to learn and review the learning process to control the outcomes. Machines support the computational capacity and the implementation of the solutions, and machines can learn based on the analytics work. Artificial intelligence focuses on the simulation of natural intelligence, taking as a paradigm human intelligence and behavior (Turing, 1950). It has succeeded in developing methods, tools, and concepts that allow machines to accomplish tasks that appear to require intelligence if humans do them, such as recognizing patterns in numbers, text, or pictures. Analytics knowledge creation is supported by machine learning/data mining techniques, deep learning, and new technologies for inference and visualization. These techniques and methods provide value to the organization in improving the quality of thinking and solving problems. Whether or not they equate to human intelligence is a debate that we do not have space to engage in here.

Nowadays, the development of analytics of things and more automated experiences drive to the search of sentient experiences in machines: "There is no reason in principle that robot designers cannot achieve the same types of combinatorial codes and thus increase the range of stimuli that can be sensed and responded to" (Pfaff, 2016). Moreover, with an approach not based on the modes of apprehension as Zubiri (1999) suggested, Husain (2017) in *The Sentient Machine* indicates the advances of artificial intelligence and the expected evolution of the Internet of Things (IoT) indicating that even though sentient machine is not close yet, the advances in developing artificial intelligence overcoming some of the barriers of human intelligence will bring sentient machines, learning from the world contact and perception by senses.

Competencies to Deal with Data

These competencies are about managing data to create usable data. Dealing with data is almost everything in analytics work. What happens is that data tasks and actions are present along with several steps of the analytics process. These range from planning the data to acquire or collect, through structuring and performing the task of gathering the data, to managing repositories, cleaning and controlling the population of data, transformation, and creation of variables and fields in data repositories. It may be surprising to the reader that this section is so short. However, as dealing with data is so central, the competencies here are very tightly integrated with those for managing and developing the analytics process and workflow. Additional detail will, therefore, be found in the following two sections. The key concept is that the competence needed by analytics professionals is beyond merely organizing/cleaning data: it is about creating data sets/tables that contain the data, structures, and variables that will feed the modeling process.

Competencies to Manage Analytics Knowledge

These competencies are about designing and maintaining the analytics knowledge management process. The use of the analytics process itself depends on how to design the process of being a smarter organization. In the adaptation of organizations and society to changes, data are one of the main resources, and creation of knowledge from data is one of the paths to follow to develop appropriate actions. The adaptation of the analytics knowledge management process and its subprocesses offers a guide to tackle the issue of achieving the adequate use of analytics knowledge. These subprocesses are analytics knowledge creation, analytics knowledge storage and retrieval, analytics knowledge transfer, and analytics knowledge application.

Analytics Knowledge Creation

The concept of creating analytics knowledge is observed in at least three ways: one is the use of analytics to create domain knowledge, another is to create analytics knowledge for the development of new technologies, and the third is the creation of knowledge about the use/adoption of analytics in the organization.

Analytics Knowledge Storage and Retrieval

Analytics knowledge storage and retrieval refers to the need to organize and manage the analytics experiences included in organizational memories. The rationale for it is that analytics knowledge is created and at the same time all too easily forgotten, but new users of analytics need to learn from the experiences of others in order to avoid a longer and sometimes tortuous learning curve. The overall purpose is to shorten the time span of moving from data to results and reports/evaluations.

Analytics Knowledge Transfer

Analytics knowledge transfer is an organizational process that goes beyond simply having the communication capacity within and outside the organization. Barriers to good communication and communication flow (Argyris, 1994) are related to defensive positions that undermine the capacity to learn and discover the real problems, superficiality in the problem definition and solutions, and almost to a kind of blindness that reduces awareness about the causes of the problem and its potential solutions.

This kind of barrier creates various limitations in the learning process and adaptation to changes: fear, and at the end a half-truth related to the problem to solve, lack of capacity to dig deeper into the issues and the potential to find solutions. Regarding these points in the current environment, organizations need more people motivated to think and formulate valuable input in the improvement process. Analytics introduction requires people inside the organizations who act as leaders of change and adoption of the analytics process. This is not a role that can be left solely to the data scientist/analytics professional.

Rodriguez and Edwards (2012) indicate that communication of an analytics process, in their case risk management, has to be improved regarding the flow and integration of policies in order to improve the board's understanding. Communication is the basis of analytics knowledge sharing and the analytics knowledge management program. Additionally, the search for effective methods to communicate analytics to the board is fundamental to the decision-making process. As Samoff and Stromquist (2001, p. 637) pointed out, "Decision-makers have very short attention spans, and they are unwilling or unlikely to read more than a few sentences on a topic. If so, for knowledge to be useful it must be presented succinctly." This view is complemented by Uzzi and Lancaster (2003, p. 395), who observed: "learning, like knowledge transfer, is a function of the type of relationship that links actors."

Analytics Knowledge Application

Alavi and Leidner (2001) indicate that knowledge application is associated with competitive advantage development and for that, there are three mechanisms to create capabilities: directives, organizational routines, and self-contained task teams. Analytics knowledge application is associated with the definition of goals, the day-to-day operations of teams and the organization as a whole. Analytics knowledge provides access to and updates of directives and supports the automation of routines such as reporting and the creation of operational systems such as credit scoring in the financial services sphere.

Competencies to Develop the Analytics Process and Workflow

These are the competencies needed to analyze, organize, develop, and put into action the analytics process and workflow using the appropriate technology, techniques, models, and methods.

Analytics Process

This section begins with a summarized view of the analytics process and how it can be materialized through the use of technology and the creation of workflows to support business processes. The competencies are associated with the possibility to answer questions using the analytics process and the design of the analytics process. The summary of the analytics process comes from Rodriguez (2017) and is illustrated using a crime dataset from Chicago (Figure 8.2).

This dataset indicates that crime is seasonal and that the number of reported crimes in Chicago is showing a downward trend. Analytics professionals need to answer questions at each of six steps:

1. Problem definition, delimitation, the definition of scope through the needs of the business. In the section on competencies for identifying the problem to study, we introduced the main theoretical ideas. An analytics professional needs to answer questions about roots of crimes, types of crimes, trends, seasonality, causes for trends, and so on, the questions ranging from general to more specific. The questions define the variables to use later in data selection/gathering and analysis. The questions go further: from identifying seasonality in crime through the possible influence of public budgeting, to benchmarking or even turning into intelligent government controlling some key performance indicators.

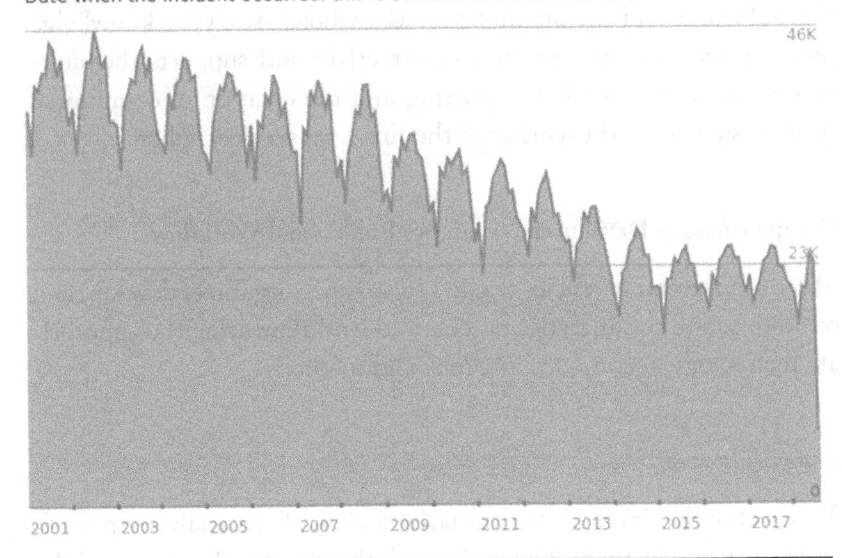

Figure 8.2 Time series of the crimes in Chicago. (Source: https://data.cityofchicago.org/Public-Safety/Crimes-2001-to-present-Dashboard/5cd6-ry5g)

2. Data management: In data gathering, analytics professionals need to deal with issues of storage, format, access to, organization of the data, cleaning data, and so on. In the crime data used in Figure 8.2, there are many sources. Figure 8.2 is a representation of an aggregated dataset that combines several variables of crime. The skill is in the organization and ability to create usable tables and data sets. Given that data are the realization of variables, data organization is clearly a step that is possible to perform once the analytics professional knows what to do to solve the problem. There is a back and forth process from problem definition and models to their use with usable data.
3. Model management: The problem and the data help drive the modeling process. However, the modeling step sometimes shows the need for new data or new variables or further clarification of the problem. For example, a time series model for forecasting crime numbers is important but probably more important in the mind of the analytics professional is to create

the description of the band (interval) of control of the crime indicator and the answers to why, what to do, and how? Why is there less reported crime in winter? What is the impact on the police department plan? How to implement steps of control of the indicator based on what has been observed?

4. Development of understanding and meaning: After the models come the results, and the analytics professional has to review them, contextualize and interpret them appropriately. The creation of meaning is the source of Know-How. This enables embedding the analytics knowledge in the business processes to develop actions of improvement/control.
5. Analytics knowledge sharing and transfer: Designing the means to share the analytics knowledge is crucial, given that the analytics work affects several stakeholders, is interdisciplinary and involves multiple teams in organizations. For instance, visualization and analysis of crime data can flow to many areas inside all the departments related to the criminal justice system.
6. Application, actions and business processes under a permanent plan. In this example of Chicago's crime statistics, the actions will come from comparing the city with other cities, to slice and dice the problem and its data, to input the definition of policies, the assignment of resources, etc.

However, the capacity of designing the analytics process is not just a list of steps to follow. It is a skill that analytics professionals need to have. Thus, the design of the process is not just to have the steps in the mind of each data analyst. It is the coordination of the "orchestra" of the team members of the analytics process development. The analytics process is not a series of independent steps; it is an interconnected set of steps that go through continuous feedback to (and from) the previous steps and iterations. For instance with the Chicago crime dataset, the deeper the analysis goes, the more questions emerge, and new possible methods to identify problem/solutions are required.

The analytics process is the subject of the analytics knowledge management process that requires the understanding of concepts that are connected to analytics workflow that comprise technology, techniques, models, and methods.

Analytics Workflow: Technology, Techniques, Models, and Methods

The analytics process must be converted into an analytics workflow to operationalize the steps of development. The concept of workflow represents the use of an organized/systematic set of activities, resources, and capabilities to accomplish what the analytics process requires. The capabilities of development are associated with technology, technique, models, and methods. The analytics process as a whole is a technology. The analytics process is systematically organized in order to perform activities and provide solutions. Cardwell (1994, p. 490) pointed out the ideas that apply to the analytics process as a technology "At the heart of technology lies the ability to recognize a human need, or desire (actual or potential) and then to devise a means—an invention or a new design- to satisfy it economically." The analytics process requires an analytics platform (technology) to develop it, which includes techniques, models, and methods to create analytics knowledge. The analytics platforms can have many tools such as data/text mining, optimization, and simulation, but the skills of the analytics professional are mainly related to organizing and using the tools according to the problem to solve.

There are many technology support possibilities: the number of algorithms for solving problems is growing at a very fast pace, there are thousands of algorithms in R and Python, and Spark big data solutions using machine learning are growing rapidly as well. Nowadays, the limitation for solving an important number of problems in organizations is not the technology. The issues are in dealing with the limitations for appropriate use of the technology, in the capacity to build the process/workflow for developing analytics, in interpreting outcomes, in explaining what algorithms have done, and in the possibility to put the results in context to create meaning for the stakeholders. In many cases, technology access is possible, but users' preparation to take advantages of the technology is limited.

One example is with applications such as spreadsheets. The original VisiCalc, created by Dan Bricklin and Bob Frankston in 1979, had most of the features that people use day-to-day in the current work environment. However, some computational applications (software) are many times more powerful than VisiCalc, yet the use of these new capabilities is very low. The problems exist, and some techniques have been developed, but the solutions have not been reached. The limited use of new capabilities can represent a risk, both operational and strategic, for organizations.

A technique, as Cardwell (1994, p. 6) said, is an "assemblage of crafts and skills that for so long met the needs of society" that in this context of analytics is the process of how analytics professionals perform actions, develop models, and tools to obtain the results of the analytics process. Models are abstractions of reality. These models can be qualitative, quantitative, graphical, etc., and are converted into tools for understanding data. The model will be composed of the selected parameters that are used for describing the particular application of the technique to the data. The difference with the technique is just in the scope of the steps and at the level where they are used in a work or research setting. In the context of analytics, the model could be a regression analysis and the methods can be associated with dealing with variable interactions, outliers, assumptions, etc. Methods refer to the use of specific steps and algorithms to obtain a clear understanding of models. For example, in the case of multiple regression methods such as forward, backward and stepwise, help to select the variables to use in the analytics process and implementation.

Analytics professionals require competencies to use technology, techniques, models, and methods as an integrated whole, to make the most of the available capabilities. The limited use of data analytics capabilities might arise from: lack of understanding of required capabilities, reduced development of capabilities, lack of human expertise, lack of understanding of analytical tools, poor use of the technological tools (as mentioned above), or poor communication capabilities in the organization. In many cases, the inappropriate use of new technology and analytics knowledge in organizations reduces the ability to search for solutions to support multidisciplinary and interdepartmental views of problem-solving approaches. In the end, the organization's requirements to successfully exploit analytics work are (Barton & Court, 2012):

- Choose the right data
- Source data creatively
- Get the necessary IT support
- Build models that predict and optimize business outcomes

- Transform the company's capabilities
- Develop business-relevant analytics that can be put to use
- Embed analytics into simple tools for the front lines
- Develop capabilities to exploit big data

The above organization's requirements are translated into the analytics platform in the form of a workflow. Assuming the data is already usable, the IT support requires skills associated with the analytics platform to work with and how the platform can be controlled and used for embedding results and process in the business process. For example, the creation of the workflow will define the PMML (Predictive Model Markup Language) to input the workflow into other support systems of business processes. We use a KNIME (n.d.) workflow example. KNIME (n.d.) is an analytics platform that has many powerful means to illustrate the organization of the analytics process in practice. The workflow diagram (Figure 8.3) represents areas that require the attention/skills of analytics professionals. The problem is to predict the answer to a promotion of a product in a banking organization.

Figure 8.3 shows how from descriptive analytics it is possible to obtain predictive analytics results. The workflow allows the comparison of different model structures and the definition of the results for classification purpose. In the analytics workflow the aspect of

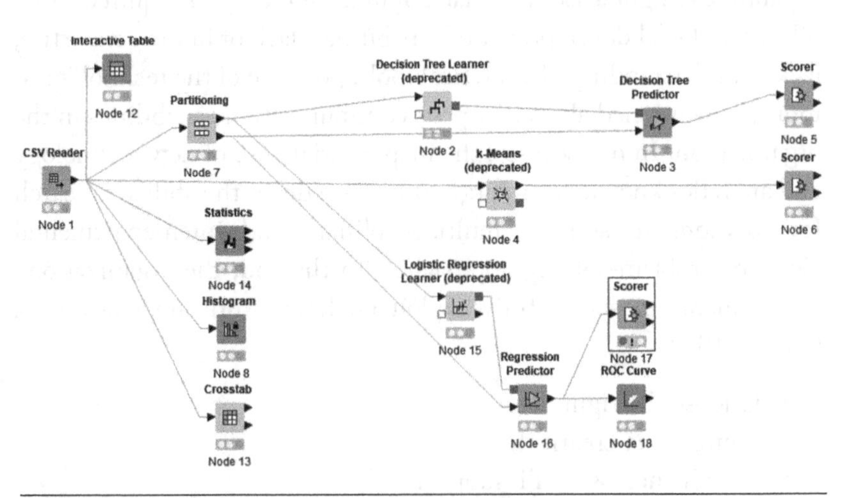

Figure 8.3 Simple model training for classification in bank promotion. (Created by authors using KNIME.)

evaluating the accuracy of models is crucial for the business process understanding. However, the analytics professional will continue digging deeper in the solutions to move from deterministic and static views in the analytics process to stochastic, interactive and dynamic treatment of the analytics process, including simulations, and the use of internet (analytics of things) as means to connect more stakeholders and solutions of a variety of problems with similar capability requirements.

Competencies for the Adoption of the Analytics Process

The competencies here are to be learners and teachers, connecting theory and practice.

Analytics professionals require competencies to reach the next layer of the analytics process, which is adoption. What organizations demand of analytics professionals is to be able to convert the outcomes of technology, techniques, models, and methods into solutions and improvements in the business processes. The requirement of rational and intuitional competencies is part of the body of the structure needed to develop analytics in organizations. On the one hand, analytics professionals need to be prepared to be learners, teachers, and implementers. On the other hand, they have to be able to work at the individual level on the development of reasoning capacity and the possibility of operationalizing the results and process of creating knowledge from data.

The movement from the analytics process and workflow to actions is equivalent to passing from a focus on Know-What and Know-Why, to a focus on Know-How. Organizations are asking how to implement solutions. We identify implementation of solutions with the full cycle of analytics adoption. New analytics professionals cannot stop with the knowledge of tools and methods but need to continue supporting the implementation and review of the solutions, especially if those solutions are delivered as "self-service" analytics. The concept of simultaneously being learners and teachers is related to what Richard Feynman (Nobel Physics Laureate, who lived 1918–1988) suggested for the learning process. Feynman fostered students to understand the concepts by trying to mimic professors explaining what a concept or solution to a problem could be, at the same time expressing the ideas of the solutions in a simplified way; for analytics, this means the presentation of complex problems and solutions using simple explanations.

In analytics work, the development of the competencies is related to the selected problem that we are interested in solving, just as in thinking we need to transfer the knowledge using the clearest way to explain to others the methods and results. In this regard, it is required to do research about the relationships among explanations, understanding and knowledge transfer. There is an ongoing search in organizations for clarifying the steps in analytics knowledge creation. More than ever, analytics professionals need the capacity to clarify and explain their work in order to make it possible to implement it. Analytics professionals, therefore, need to maintain the search for better explanations of steps where some aspects are not clear.

The concept of being implementers and action-oriented can be found in the work of Argyris (1994) and Pfeffer and Sutton (1999). Argyris (1994) introduced the concept of managing "defensive reasoning" and the way to promote a search for solutions-actions to answer the needs of the organization. The main point to bring to this presentation is the concept of double-loop learning which is crucial to add value in the analytics process. Argyris indicates that in a double-loop learning there are questions about the facts and about the reasons and motives behind the facts. In an analytics process, it is required the understanding of the results and circumstances around those results in the way that the phenomena-organization has been affected. He also pointed out, referring to the communication practices of managers and the reactions of the employees (Argyris, 1994, p. 77): "They do not encourage individual accountability. And they do not surface the kinds of deep and potentially threatening or embarrassing information that can motivate learning and produce real change." And, nowadays analytics professionals need to promote the development of work that is not only technical and facts search, but also about reasons of results and implementation of possible solutions. Argyris (1994, p. 85) concluded in his study: "Today, facing competitive pressures an earlier generation could hardly have imagined, managers need employees who think constantly and creatively about the needs of the organization. They need employees with as much intrinsic motivation and as deep a sense of organizational stewardship as any company executive."

The analytics professionals need to have an ongoing search for competencies that are in the areas of implementation, generating actions and promoting the solutions. There are traps in having analytics knowledge in the sense of what Pfeffer and Sutton (1999) called the "Smart-Talk Trap" in the sense that even though the know-how is in place, the implementation requires actions to enhance organization's performance. Pfeffer and Sutton (1999, p.21) pointed out,

> In today's business world, there's no shortage of know-how. When companies get into trouble, their executives have vast resources at their disposal: their own experiences, colleagues' ideas, reams of computer-generated data, thousands of publications, and consultants armed with the latest managerial concepts and tools. But all too often, even with all that knowledge floating around, companies are plagued with inertia that comes from knowing too much and doing too little—a phenomenon the authors call the knowing-doing gap.

Three domains of understanding support the analytics process: statistical, computational and business. According to Garfield and Ben-Zvi (2008), there are three main aspects that statistical capabilities involve: statistical literacy, statistical reasoning, and statistical thinking. The adoption of the analytics process requires a similar preparation of analytics professionals.

Analytics literacy covers the technology, techniques, models, and methods that exist in analytics and that use statistics and computational components. This is equivalent to what is represented by Garfield and Ben-Zvi (2008) as statistical literacy "[it] involves understanding and using the basic language and tools of statistics: knowing what basic statistical terms mean, understanding the use of simple statistical symbols, and recognizing and being able to interpret different representations of data."

Analytics reasoning and analytics thinking represent the capacity to go further than the basic knowledge of algorithms, to reach the level of creating solutions, analytics knowledge, and the possibility to learn and develop new answers (knowledge and pattern recognition) to an organization's problems looking for developing actions. Figure 8.4, from Edwards and Rodriguez (2016), shows how these come together in an analytics project.

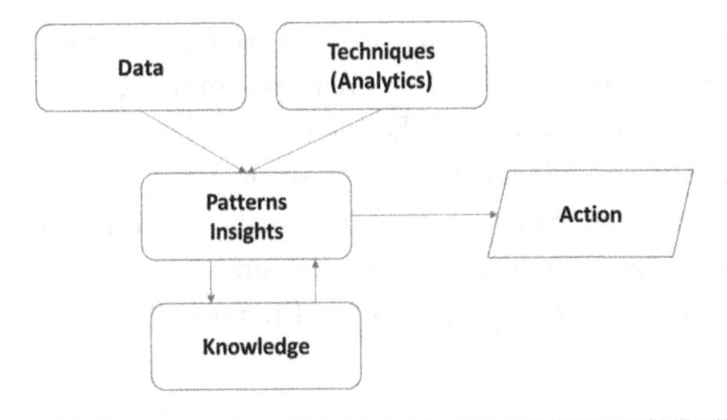

Figure 8.4 Interactions between data, analytics, and human knowledge (Edwards & Rodriguez, 2016).

We propose to extend the concepts from Garfield and Ben-Zvi (2008) from the statistical component of analytics to the computational part as well. It is possible to consider analytics reasoning as the analytics professional provides sense to the analytics knowledge, and analytics thinking such as the way that professionals in analytics can choose, manipulate, interpret, apply, and decide about the technology, techniques, models, and methods to use. The computational aspect of analytics adoption refers to the development of the capacity to define steps/rules/operations to process data to obtain results, or as Rowland (n.d.) defines, "A computation is an operation that begins with some initial conditions and gives an output which follows from a definite set of rules." These steps/rules are defined by the creation or use of algorithms in certain programming languages. The adoption of analytics suggests the need of competencies for creation and implementation of computational capacity in the various areas of the organization.

From the business understanding point of view, the adoption of the analytics process for the analytics professional requires involvement in the organization's innovation, the organization's control of operational and strategic risk (productivity, competitiveness, and performance improvement). In addition, it is important the understanding of the organization's adoption of other technologies such as customer/supplier relationship management and the need to analyze customer/supplier data. Analytics is part of the innovation process (Duan, Cao, & Edwards, 2018) as a two-way process because analytics in one direction support senior management to review problems considered important to study and in the other direction to define the problem and find an analytics solution.

Business understanding in analytics adoption is related to the capacity to deal with uncertainty looking for controlling strategic/ operational risk. The most common definition of operational risk, first published by the Basel Committee on Banking Supervision (2006), includes the loss, direct or indirect, that is a result of a failure in internal processes, people actions, and systems performance. Many of the risks in the operational risk categories can be related to the use of technology, techniques, models, and methods. For example, many types of risk are associated with problems that analytics people study: internal and external fraud, employee's malpractice, safety in the workplace, damage to physical assets, disruption of business and systems failures, and low quality of execution, delivery, and processes management.

Moreover, adoption of analytics is associated with the formulation/ control of business objectives, strategy and governance model. This principle is part of the process of controlling strategic risk based on analytics knowledge. Adoption of analytics starts with the definition of business objectives. Design and implementation of the strategy and governance models require the development of tools, coordination, integration, and aggregation of data, multiple information systems, and analytics technologies. Setting objectives based on experience, time series analysis, forecasting techniques, or other techniques will be related to the way that results will be measured. For example, once organizations identify Key Performance Indicators and Key Risk Indicators organizations require to continue the analytics process and revisit the results obtained in each period of measurement or observation of the metrics value.

Summary and Review

In this chapter, we have presented the required competencies for the analytics process with the purpose of enabling its adoption in organizations. These cover identifying the problem, using intelligences (plural), dealing with data, managing analytics knowledge, developing the analytics process and workflow, and securing analytics adoption and actions. The main purpose was to demonstrate that competencies are not only the appropriate uses of intelligences, analytics techniques and knowledge, but also in their implementation, in the know-how and creation actions. Charan (2007, p. 68) pointed out, "In my observation, people who create organic growth that is profitable and sustainable connect the dots sooner and are on the offensive."

Data analytics is connected to many different technological/organizational structures and purposes. However, one common point in adoption is the search for technologies that contribute to growth in organizational performance. Moreover, new technologies support not only operational processes but also the strategic processes within the organization. Thus, currently, not only is there the need for competencies to deal with new technologies, and the growing volume of data, but also to deal with the limitation of organizational capabilities to obtain knowledge from data, use the knowledge to solve problems and support complex decisions. The challenge is to put the analytics process to work. To avoid stopping at the level of particular answers and to develop the systems to maintain the answers to problems in a way that aligns with the organization's goals. In many cases, the gaps between new technology and analytics knowledge constrain the search for solutions to support multidisciplinary and interdepartmental views of problem-solving approaches. Thus, the preparation of an analytics professional needs to be on the layers of nurturing the intelligences (strategic, collective, artificial, and sentient), and acquiring knowledge of new technologies, techniques, and methods, but mainly in the support of the operations, the capacity to create know-how, implementation, and full adoption of the analytics process.

References

Alavi, M., & Leidner, D. (2001). Review: Knowledge management and knowledge management systems: Conceptual foundations and research issues. *MIS Quarterly, 25*(1), 107–136.

APEC. (2017). Data science and analytics skills shortage: Equipping the APEC workforce with the competencies demanded by employers. Retrieved from https://www.apec.org/Publications/2017/11/Data-Science-and-Analytics-Skills-Shortage

Argyris, C. (1994, July–August). Good communication that blocks learning. *Harvard Business Review*. Retrieved from https://hbr.org/1994/07/good-communication-that-blocks-learning

Barton, D., & Court, D. (2012, October). Making advanced analytics work for you. *Harvard Business Review*. Retrieved from https://hbr.org/2012/10/making-advanced-analytics-work-for-you

Basel Committee on Banking Supervision. (2006). International Convergence of Capital Measurement and Capital Standards, Bank for International Settlements. Retrieved from https://www.bis.org/publ/bcbs128.htm

Cardwell, D. (1994). *The Fontana History of Technology*. London: Fontana Press/Harper Collins.

Charan, R. (2007). *Know-How*. New York: Crown Business.

Duan, Y., Cao, G., & Edwards, J. S. (2018). Understanding the impact of business analytics on innovation. *European Journal of Operational Research*. doi:10.1016/j.ejor.2018.06.021.

Edwards, J. S., & Rodriguez, E. (2016). Using knowledge management to give context to analytics and big data and reduce strategic risk. *Procedia Computer Science, 99*, 36–49. doi:10.1016/j.procs.2016.09.099.

Garfield, J.B., & Ben-Zvi, D. (2008). *Developing Students' Statistical Reasoning: Connecting Research and Teaching Practice*. New York: Springer.

Garud, R. (1997). On the distinction between know-how-know-why, and know-what. *Advances in Strategic Management, 14*, 81–101.

Henke, N., Bughin, J., Chui, M., Saleh, T., Wiseman, B., & Sethupathy, G. (2016). *The age of analytics: Competing in a data-driven world*. McKinsey Global Institute. Retrieved from https://www.mckinsey.com/business-functions/mckinsey-analytics/our-insights/the-age-of-analytics-competing-in-a-data-driven-world

Husain, A. (2017). *The Sentient Machine: The Coming Age of Artificial Intelligence*. New York: Scribner.

KNIME. (n.d.). *Knime*. Retrieved from https://www.knime.com

Montgomery, D., & Weinberg, C. (1981). Toward strategic intelligent systems. *The McKinsey Quarterly, 1*, 82–102.

Pfaff, D. (2016). *A Neuroscientist Looks at Robots*. Singapore: World Scientific Publishing.

Pfeffer, J., & Sutton, R. I. (1999). The smart-talk trap. *Harvard Business Review, 77*(3), 134–142.

Piaget, J. (1963). *The Psychology of Intelligence*. New York: Routledge.

Price Waterhouse Coopers and the Business-Higher Education Forum. (2017). *Investing in America's data science and analytics talent*. Retrieved from http://www.bhef.com/sites/default/files/bhef_2017_investing_in_dsa.pdf

Rodriguez, E. (Editor). (2017). *Analytics Process: Strategic and Tactic Steps*. Boca Raton, FL: CRC Press.

Rodriguez, E., & Edwards, J. S. (2012). Transferring knowledge of risk management to the board of directors and executives. *Journal of Risk Management in Financial Institutions, 5*(2), 162–180.

Rodriguez, E., Edwards, J. S., & Facundo A. (2011). Country strategic risk and knowledge management: A proposed framework for improvement. *Proceedings European Conference Knowledge Management, ECKM, 2*, 836–846.

Rowland, T. (n.d.). Computation. *MathWorld-A Worlfram Web Resource, created by Eric W. Wesstein*. Retrieved from http://mathworld.wolfram.com/Computation.html

Samoff, J., & Stromquist, N. (2001). Managing knowledge and storing wisdom? New forms of foreign aid? *Development and Change, 32*(4), 631–656. doi:10.1111/1467-7660.00220.

Service, R. (2006). The development of strategic intelligence: A managerial perspective, *International Journal of Management, 23*(1), 61–77.

Turing, A. (1950). Computing machinery and intelligence. *Mind, 59*(236), 433–460.

Uzzi, B., & Lancaster, R. (2003). Relational embeddedness and learning: The case of bank loan managers and their clients. *Management Science, 49*(4), 383–399.

Zubiri, X. (1999). *Sentient Intelligence.* (T. Fowler, Trans.). Washington, DC: Xavier Zubiri Foundation of North America.

9

Business Analytics

A Course Design

KATHERINE LEAMING GOLDBERG

Washington College

Contents

Introduction

As institutions move toward adding programs in data analytics, in some cases the introduction of a single course is a good first step. The purpose of this chapter is to share the outline of such an undergraduate introductory course in analytics as taught through the lens of Business Management, including details on content taught as well as specific examples of the homework assignments and weekly readings. Entitled "Business Analytics," this course serves as an elective for the Business Management major and minor as well as the interdisciplinary Information Systems minor. Management Information Systems and a one-semester statistics course are required prerequisites, and most students are juniors or seniors majoring in Business Management or Computer Science, or minoring in Information Systems. The text used in this course is "Business Analytics: An Introduction," edited by Jay Liebowitz. An essential element to ensure sustainability for this course lies in keeping faculty skills current, so I will include resources that can be helpful in this regard.

Learning goals of this course center on the collection and analysis of data as a means of transforming them into actionable knowledge within the context of Business Management. The learning outcomes of this course are as follows:

- Managerial knowledge
- Critical thinking
- Quantitative analysis
- Communications skills
- Global perspectives
- Collaboration skills
- Ethical awareness

A week-by-week outline of the class including the materials covered, homework assignments, and readings is provided in Appendix A.

Overview of the Course

For some students, this might be their first real foray into analytics, and it can be a bit intimidating. I first start with definitions, and then provide an historical context, explaining the evolution of modern statistics and computer science. Next, I provide examples in the application of Business Management that the students are likely to have already experienced such as Netflix, Amazon, Spotify, and Facebook. We then have discussions about data-driven decisions and decision bias. I give examples of different decision biases that people might fall into along with examples so that students can understand how difficult it can be to make decisions.

Because Microsoft Office Excel is still a widely used data analytics tool and because students have already had multiple exposures to it, using Pivot Tables in Microsoft Excel is the first assignment given to the students. This allows students to gain a simple understanding of the types of basic questions that can be answered by aggregating the data. I begin with this assignment because it does not require complicated math or algorithms, and it allows me to assess each student's comfort level with Excel, which they will need for many of the assignments. The step-by-step details of assignments are provided later in the chapter. This assignment will help start the students with their quantitative analysis and critical thinking learning outcomes.

During the second week, I cover why analytics are needed and why analytics are important in a business by leading students to consider examples of how decision-makers use analytics to increase profits, reduce costs, retain customers, and decide what products or markets to use. I share white papers about analytics maturity models and explain how organizations move through different stages of analytic maturity. I also cover the different levels of analytics: descriptive, predictive, prescriptive, and text. Lastly, I explain how data analysts approach analytical problems or projects using the CRISP-DM methodology. CRISP-DM stands for Cross-Industry Standard Process for Data Mining; more information about this methodology and process model can be found at the IBM Knowledge Center (www.ibm.com/support/knowledgecenter/en/SS3RA7_15.0.0/com.ibm.spss.crispdm.help/crisp_overview.htm).

The next assignment uses SAP Business Objects to connect the data into Microsoft Excel. SAP is a global software company providing transactional and analytical software products to support enterprise management and decision-making. Membership in the SAP University Alliances program provides colleges and universities with access to the SAP software required for these exercises as well as detailed step-by-step curricula. Membership also provides faculty with access to training workshops and an online learning community for developing and maintaining skills with the various software products provided by the program. Information about the SAP University Alliances program can be found at www.sap.com/training-certification/university-alliances.html#faculty. By connecting data in this assignment, students build on their knowledge from the Pivot Table assignment, but in more depth because there are more filter options and aggregation options that cannot be done simply by using Pivot Tables.

The next topic is dashboards and visualizations in which students learn how to share with their readers the data they have aggregated. I show the students examples of visualizations that are not helpful as well as best practices. The students are asked to interpret many different dashboard elements and uncover what works and what doesn't. By being asked to interpret dashboards and visualizations created by others, students also learn how other people might interpret student outputs.

The next project is to use Tableau to create 12 visualizations to practice each option in Tableau. The students then build their own dashboards using the visual elements they created. They demonstrate critical thinking by selecting the visualizations that best convey information to the reader and writing a paper detailing why they selected each visualization.

Up until this point, the students have been dealing with very simple, very small data sets. In the business world, it is more likely that a data analyst would encounter larger data sets in terms of both rows and columns. What happens when you get larger data and the business consumer has questions that cannot simply be answered by providing simple descriptive statistics or visualizations? We talk about the definitions of data mining as well as the needs of data warehouses. I provide definitions of the data mining techniques of clustering, prediction, classification, and association. We walk through definitions of each technique and why they are necessary.

Next, I dive into each of the data mining techniques beginning with clusters because it is the easiest to understand and because they had just learned visualization methods. The students learn how to build clusters using both Tableau and SAP Predictive Analytics. In Tableau, there are options to build clusters under the analysis toolbar. Tableau assigns each data element to a cluster, and then you can use the data visualization tools to create representations of the clusters. Within SAP, there are options in Predictive Analytics that lead the user through creating clusters, and then multiple visualizations are used to show the clusters. A data set about retail stores is used to uncover characteristics such as sales turnover, profit margin, and store size. During the lectures, I introduce the concepts and definitions of big data and its application in business and cover what unstructured data encompasses.

Now that the students have successfully completed a simple technique of data mining, they are ready for Predictive Analytics, learning about the difference between supervised and unsupervised algorithms as well as classification algorithms and definition and applications.

In order to learn classification, I use an excellent SAP exercise (Introduction to Next-Generation Business Intelligence with Automated Analytics Mode in SAP Predictive Analytics Case Study) that shows the students how to build decision trees and interpret the results. When the students complete this exercise and answer questions, they are demonstrating managerial knowledge as well as quantitative analysis.

By this time in the class, we are close to the midpoint, and the students should have an in-depth knowledge of the basic approaches of data analytics. Up until now, the students should see that they could use fairly simple tools such as Microsoft Excel, Tableau, and SAP to perform the most basic analytics up to predictive analytics.

Because the data aren't always small and simple, the students need to understand how businesses might use relational databases and how the data are related to each other. What happens when the data need to be updated? How can a user access the data? What is structured query language, and how does it work? Students practice writing SQL code by completing an online SQL tutorial that is prepopulated with data; the students simply write code to answer questions. The tutorial can be found at www.w3schools.com/sql/default.asp. I also show students examples of how data tables relate to each other. Once they have these basics, the natural question is how do data warehouses relate?

What happens when the data grow and grow and grow? How are businesses handling big data? What is Hadoop, and what companies use it? I expose students to the broad concepts of Hadoop. What is in-memory processing? What is SAP HANA? Videos are used to introduce the concept; then students further explore through a hosted exercise available from the SAP University Alliance covering big data using SAP ERP on HANA.

By now, the students should have a solid understanding of the basic data analytic methods and so the next natural progression is to explain how to perform text analytics. We talk about the origin of the text data, how to prepare text data and the analytics that can be used such as clusters, decision trees, and visualizations in the form of word clouds. Next, we discuss how businesses use analytics to further their missions and the many applications of it. I have developed a group assignment using text analytics in which students perform a Web scrape of reviews of local restaurants from the TripAdvisor website. I assign each team a "target" restaurant that is client who has hired their consulting firm to perform an analysis. Students export their data into an Excel spreadsheet and perform a sentiment analysis on these data. Each member of the team reviews a different restaurant and then they compare their findings and make recommendations to their customer. This satisfies their learning outcome of demonstrating collaboration skills.

Up until now, the students have been dealing with fairly "math-based" and concrete algorithms. Next, I introduce a more abstract concept of machine learning. I walk them through the definitions as well as applications. I offer examples of companies that are using machine learning to build their brands and sell their products. Next, I introduce neural networks and explain how they work. I do provide backgrounds and explanations on how neural networks work, but only for purposes of exposure to the complex ideas in machine learning.

The students complete their first machine learning assignment using a product called BigML, Inc., which is a Web-based and free product available to students. This company has an academic agreement and program that can be found here: https://bigml.com/education. Later in this chapter, I provide an example of an assignment using BigML, Inc., to expose students to machine learning. The learning outcomes after the completion of this assignment are managerial knowledge, critical thinking, and communication skills.

Another very complex topic in data analytics is geographic mapping. I use a homework assignment that combines SAP Lumira and ArcGIS to teach the students how to build maps and perform analysis of data using the maps. ArcGIS is a mapping and analytics product offered by ESRI; you can learn more about it at www.esri.com/en-us/arcgis/about-arcgis/overview. The students answer questions throughout the assignment to demonstrate their knowledge. Because I am an ArcGIS user, I have a subscription to a quarterly magazine called *ESRI News*. I assign each student an article from *ESRI News* and asked them to prepare a 20 min presentation to the class about the article and how data analytics was used to solve a particular problem. This assignment is used to ensure the students demonstrate communication skills as well as global perspectives since many of the articles are about international communities.

Social media is something most students are very familiar with, but they often do not know the history of how each social media platform was built and grew. They also do not always understand the application of business for each social media. I go through the major platforms of Facebook, Twitter, LinkedIn, and YouTube and talk about the data analytics that might be applied to each. I encourage the students to think about their own presence on each of the different sites in terms of how their data are analyzed and consumed.

The natural next step about learning about how data are used by companies is to understand privacy and ethical concerns. I explain the various data regulations in the United States, Europe, Asia, and South America. We also discuss the definition of ethics and different dilemmas that arise in data analytics. Through readings and conversations, the students should be able to demonstrate an ethical awareness associated with data analytics.

The last topic discussed covers the employment trends in the industry with examples of job titles, descriptions, and types of companies. We go through examples of what kinds of companies might look for certain skills. We also talk about the free resources that students can access to hone their skills further or learn new skills.

In the next section, I will provide details about three assignments spanning from early introduction of concepts in the Excel Pivot Table exercise, to understanding visualization using Tableau, and ending with a machine learning and classification assignment using BigML, Inc., that requires the integration of several skills.

First Assignment: Pivot Table

The course begins with definitions of analytics and the value it brings to businesses. The prerequisite is a course in Management Information Systems, so the students should already have an appreciation of the value of using software systems to solve business problems. In addition, in the prerequisite class, they have been exposed to database concepts using Microsoft Access, and they have acquired knowledge of business processes and the transactional data created by completing the Order Fulfillment Cash-to-Cash process using SAP S/4Hana Enterprise software. They have also been introduced to business analytics concepts, including a high-level introduction to visualization and big data with respect to the Internet of Things.

The purpose of the Excel Pivot Table assignment is to guide the students through learning how to answer questions about a data set that cannot be answered by simply sorting or filtering the spreadsheet. I use a sample data set about sales of camping equipment in

	A	B	C	D	E	F	G
1	Transaction	Location	Date	Item	Wholesale Price	Units Sol	Total Sale
2	20100000	California	29-May	Tent	$199.00	2	$398.00
3	20100001	Washington	17-May	Headlamp	$39.99	2	$79.98
4	20100002	Washington	20-May	Sleeping Bag	$58.50	1	$58.50
5	20100003	Washington	14-May	Headlamp	$39.99	1	$39.99
6	20100004	California	7-May	Tent	$199.00	3	$597.00
7	20100005	Oregon	22-May	Backpack	$98.77	1	$98.77
8	20100006	Oregon	6-May	Backpack	$98.77	1	$98.77
9	20100007	California	2-May	Car Rack	$415.75	2	$831.50
10	20100008	California	6-May	Backpack	$180.50	1	$180.50
11	20100009	California	5-May	Car Rack	$415.75	1	$415.75
12	20100010	California	13-May	Backpack	$220.30	1	$220.30
13	20100011	California	5-May	Headlamp	$39.99	4	$159.96
14	20100012	Oregon	28-May	Backpack	$98.77	1	$98.77
15	20100013	Oregon	5-May	Car Rack	$415.75	1	$415.75
16	20100014	Oregon	17-May	Backpack	$98.77	1	$98.77
17	20100015	Oregon	21-May	Backpack	$98.77	2	$197.54

Figure 9.1 Excerpt from raw data set for Pivot Table problem.

the Northwest United States (see Figure 9.1), but any data set can be used for this assignment as long as it contains at least 100 rows and 5 columns. The data should have a few different categorical fields (store name, region, or product name) and a few continuous fields (cost, units, or profit). It is important to instruct the students to remove any blank rows or columns; in addition, the data should not contain any existing totals or calculations. I also instruct students to be sure that the column headings make sense to them; for example, if the raw data column says "column 1," the students should change the column to an appropriate name.

Next, the students create a Pivot Table by going under the "Insert" menu and selecting "Recommended Pivot Table" (see Figure 9.2); this selection is helpful in walking the students through their first Pivot Tables. Once they are more comfortable, they can simply select "Pivot Table" to build one from scratch.

First, they are instructed to find the "Total Sum of Sales, Sum of Wholesale Price, Count of Transaction by Item". Using this

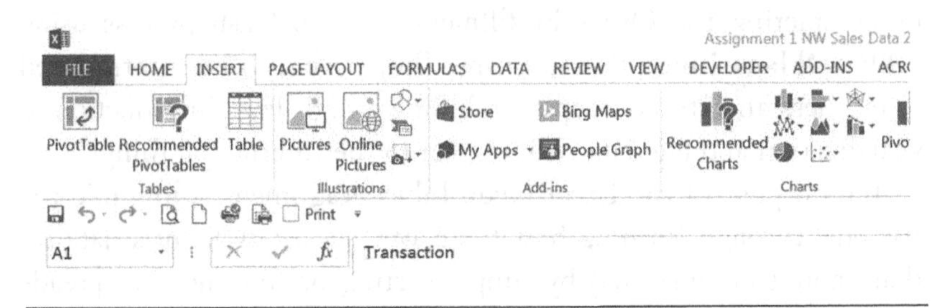

Figure 9.2 Selection of "Recommended Pivot Table."

recommended Pivot Table will help to show the students what an example looks like (see Figure 9.3).

The resulting Pivot Table will be similar to the table shown in Figure 9.4. For each product sold by the company, we can see the total sales, total sum of the wholesale price, and the counts of the sales transactions. Pivot Tables also automatically create grand totals for each column as well.

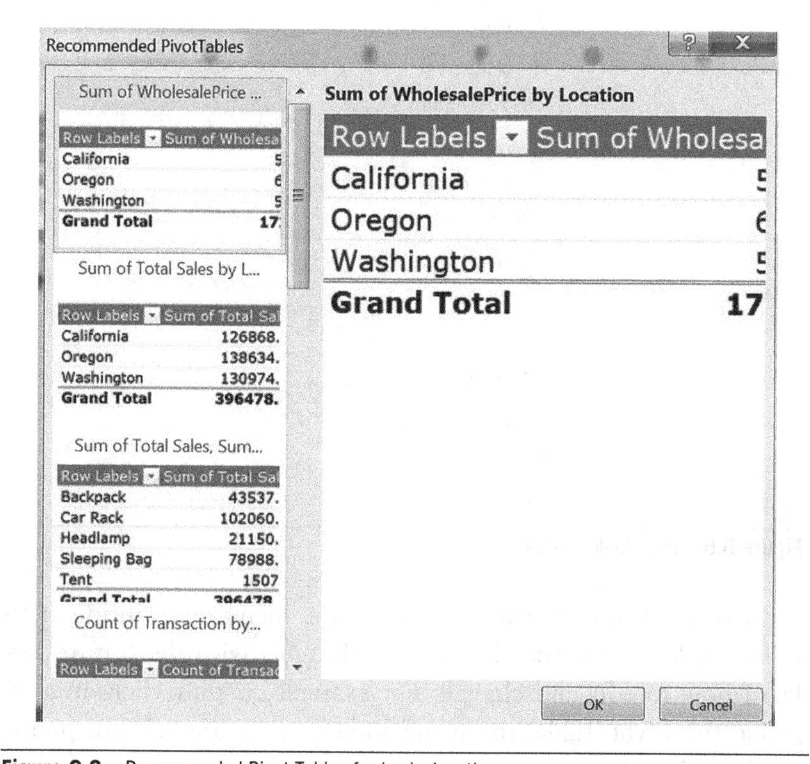

Figure 9.3 Recommended Pivot Table of price by location.

	A	B	C	D
1				
2				
3	Row Labels	Sum of Total Sales	Sum of WholesalePrice	Count of Transaction
4	Backpack	43537.03	29152.93	155
5	Car Rack	102060.55	67483.79	147
6	Headlamp	21150.94	7881.18	163
7	Sleeping Bag	78988.67	25120.75	173
8	Tent	150741	43120	150
9	**Grand Total**	**396478.19**	**172758.65**	**788**
10				
11				

Figure 9.4 Completed Pivot Table with the sums and counts for each product.

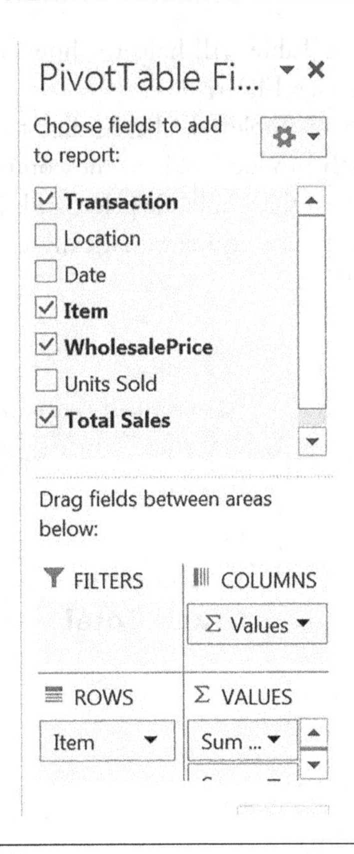

Figure 9.5 Pivot Table options.

Next, I ask the students to explore the menu items inside of the Pivot Table so that they become comfortable with the features and learn how to edit and change. For example, if they click on a cell inside the Pivot Table, the menu shown in Figure 9.5 will pop up on the right side.

The students are instructed to examine the filter options as well as the calculations so that they can see, for example, that they can perform averages rather than summations. Their assignment is to answer the questions listed in Table 9.1 in complete sentences along with their spreadsheet.

This assignment can be easily modified to add more questions. I use this as the first assignment to assess each student's level of comfort with Microsoft Excel. The final question is asking the students to use their critical thinking skills to think about the difference between totaling up all the values in a column versus performing a count of the rows.

Table 9.1 Questions for Pivot Table Assignment

a. Which location had the *highest average* sales?
b. Which item had the *lowest total of units* sold?
c. What were the *total sales of tents* in *Oregon*?
d. What is the **count of units sold** in *California*?
e. Why is there a difference between the sum of "Units Sold" and the count of "Units Sold?"

Visualization Assignment Using Tableau

There are many data visualization software packages available, but I use Tableau to teach this because educational licenses are available from Tableau and because it is widely used in the data analytics industry. Tableau educational licenses are free of charge but must be obtained from the educational outreach department with Tableau (go to www.tableau.com/academic). The software can be installed on lab computers as well as the students' machines (both Microsoft Windows and MacOS operating systems).

In addition to having access to the licensed software, the students also have access to the how-to videos and step-by-step instruction. This assignment uses the sample data for a Superstore that is automatically loaded inside of Tableau. Figure 9.6 illustrates the launch of Tableau; the Sample Data (Superstore) appears under Saved Data Sources.

Then, I direct students to a video-training website, www.tableau.com/learn/training, and ask them to go through each type of data visualization and follow the steps in each walkthrough found at https://onlinehelp.tableau.com/current/pro/desktop/en-us/dataview_examples.html. There are 12 visualizations that they are instructed to create and practice building. The 12 elements are bar chart, text table, line chart, scatter plot, highlight table (heat map), histogram, Gantt chart, pie chart, treemap, box plot, packed bubble chart, and a map. After they complete all 12, I ask them to interpret each visualization and answer questions specific to each visualization, as follows:

1. Bar chart: In the East region, which year saw the highest amount of First-Class shipping? How much was it?
2. Text table: In 2014 for ALL regions, which subcategory had the least and highest percent of sales?

Figure 9.6 Launch screen of Tableau.

3. Line: Using the final plot created, what pattern would you tell the executives about sales? Are there certain times each year when the sales are lowest? Are there seasons when sales are higher? Why do you think it would be helpful for a company to use the "forecast" feature?
4. Scatter plot: Which category has the least increase in profit as sales increase?
5. Highlight table (heat map): In your opinion, what do you think about the visualization you created? Is it helpful? What information is gathered from this?
6. Histogram (note: the screenshots in the instructions will not match your data exactly): Using the last graphic, is there a normal distribution of quantities? Why or why not?
7. Gantt chart: Using the final Gantt chart created, explain at least one insight that you can gain from this visualization. If you worked for this company, why would this be helpful?

8. Pie chart: Name at least two concerns you have about this pie chart.
9. Treemap: Using the final version of the treemap you created, what immediately sticks out to you about this information? What would you want your audience to understand from this visualization?
10. Box plot: Which region had the highest discounts for the consumer and corporate segments?
11. Packed bubble chart: Which combination of region and category had the most profit? Do you think this is a helpful chart? Explain your answer.
12. Map view: From the last map you created, which city would you estimate has the highest sales? Was this easy to determine using the visualization ONLY? (Feel free to do the Map with Pie Marks if you feel so inclined).

After they answer these questions, they are asked to create a dashboard using at least four of the elements that they created. They should pick the four elements that they think best answer questions for executives of the fictional company and then write a one- to two-page explanation of why they picked the elements and how they hoped to share information with the business leader of the company.

This assignment can be expanded by adding more questions about the visualizations they created or to ask the students to modify the visualizations beyond the instructions provided in the Tableau training. Note, the data in the sample data set are updated over time and the instructions are also changed frequently. So, it is important to update the assignment with the most recent changes in Tableau.

Machine Learning and Classification Assignment Using BigML, Inc.

BigML, Inc. (https://bigml.com) is a Web-based machine learning data analytics platform. In addition to being very user friendly and easy to quickly learn, it also contains many preloaded data sets to explore. It is free to students using their college or university email address (.edu). I encourage the students to watch an 18 min introduction video (www.youtube.com/watch?v=w0jRGVwDHn4) and also read other tutorials here: https://bigml.com/tutorials/. Once they are

Figure 9.7 Footer from the BigML, Inc. website for accessing the data sets in the gallery.

familiar with the layout of the program, they are instructed to browse through the gallery on the website and select "datasets." The Gallery is accessed from the footer of the BigML, Inc. website (see Figure 9.7).

The goal of this assignment is to empower students to interpret information from a visualization and articulate recommendations. An important decision for businesses is in customer retention and understanding behaviors that might lead to a customer leaving. Homework assignments focus on determining which customers they are going to keep using the telecom data set to analyze customer churn and explaining these results to a hypothetical audience of company executives. I will describe two of these techniques (decision tree and clusters); logistic regression analysis can also be used. For the decision tree assignment, students select three branches ending in a terminal node from the decision tree, providing the confidence as well as number of instances for each decision. For the cluster assignment, they select two clusters and identify characteristics that would be helpful for the company to understand so they can retain their customers. In their explanation, they need to provide recommendations for helping to prevent customers from leaving. This homework assignment can be expanded by asking the students to select another data set and provide an analysis.

For this exercise, the students are instructed to find "Churn in Telecom's dataset" (see Figure 9.8) and to click the "free" tag. It will say it was purchased, but because it was free, the students will not be charged. There are data sets and other items for purchase, but none are used in this assignment.

A prompt, asking if they want to "clone" the data set appears, and they should click "yes."

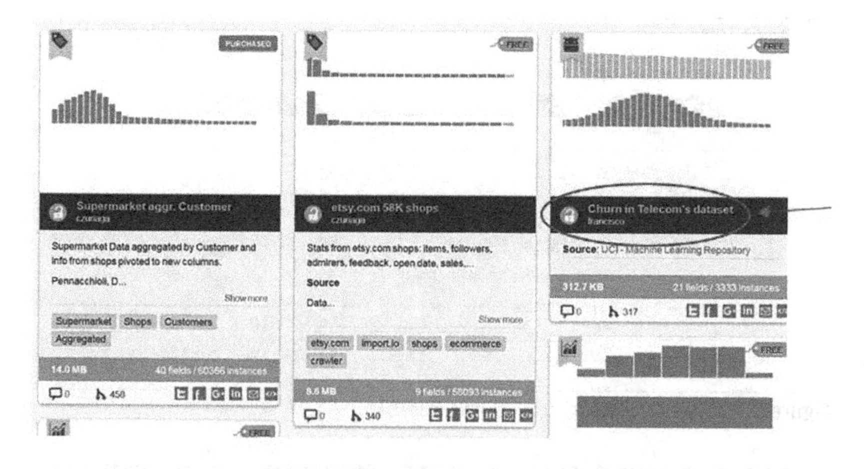

Figure 9.8 Example of the data download options from the gallery.

PROJECT: All

Sources | Datasets | Models | Clusters | Anomalies | Associations | Predictions | Tasks | Scripts

Churn in Telecom's dataset

Name		Type	Count	Missing	Errors	Histogram
state	!	ABC	3,333	0	0	
account length		123	3,333	0	0	
area code		123	3,333	0	0	
phone number	!	text	3,333	0	0	

Figure 9.9 Data set properties of each field including a histogram visualization.

Now, the students should see these data in the "data set" tab of their environment, and they can browse through the variables to inspect the types, counts, and histograms (see Figure 9.9). I always like to emphasize that this is building on the descriptive statistics that they learned earlier in the course when they learned how to create histograms. This data set does not require any cleaning or manipulation.

Next, the students should create a model by clicking on the gears icon to "configure the model" by selecting "Model" (see Figure 9.10) under Configure Supervised.

This will present a Model Configuration field. If it does not default to Churn, select Churn as the Objective Field (Figure 9.11) and then select Create Model.

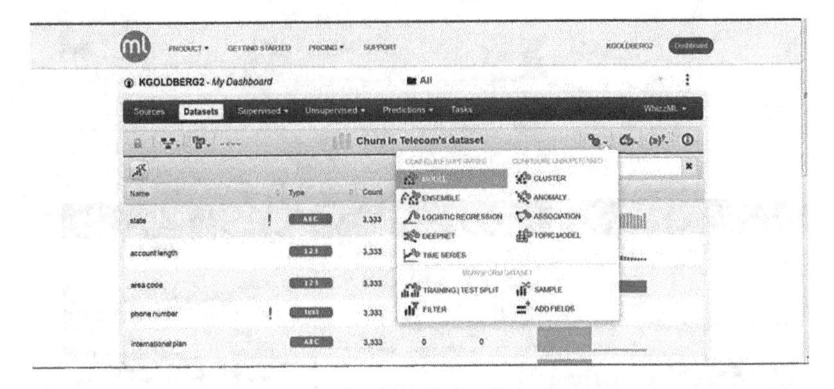

Figure 9.10 Menu options for configuring a model.

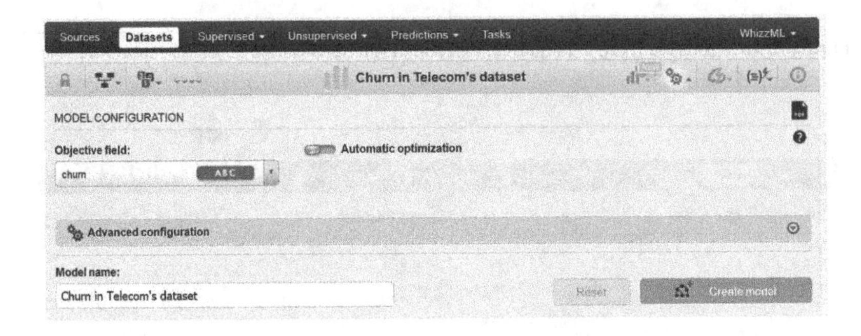

Figure 9.11 Model configuration.

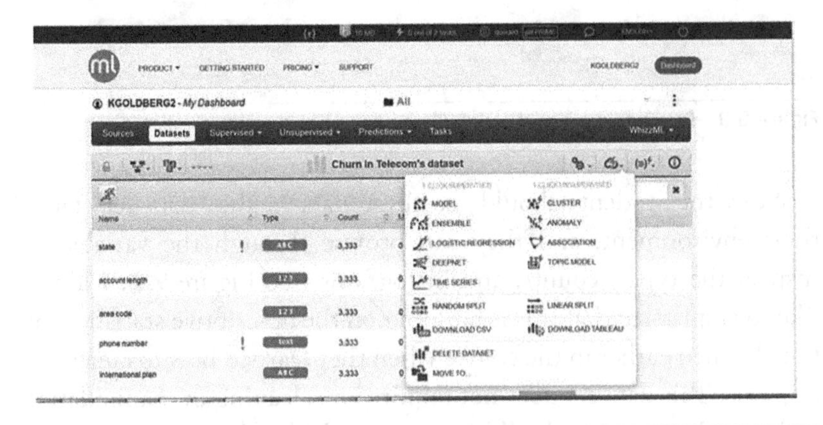

Figure 9.12 Options under the one-click model generation tool.

Now they can finish creating the model by clicking the icon of the cloud with the lightning bolt (see Figure 9.12). The first option under 1 Click Supervised, "model," will create a decision tree to predict churn.

I encourage the students to explore the decision tree created, toggle the target variable by changing it from True to False, and examine branches that end in terminal nodes (see Figure 9.13). When Churn is predicted to be False, it means that the customer is predicted to stay with the firm.

The second part of this assignment is to build clusters. Going back to the gear icon, students choose "Clusters" under "Configure Unsupervised." They then ensure that the Clustering Algorithm is set to "*K*-means" and choose the number of clusters (*K*) they would like to see (see Figure 9.14).

The results of the clusters create a visualization that allows the students to understand the characteristics for each cluster by hovering over each cluster (see Figure 9.15).

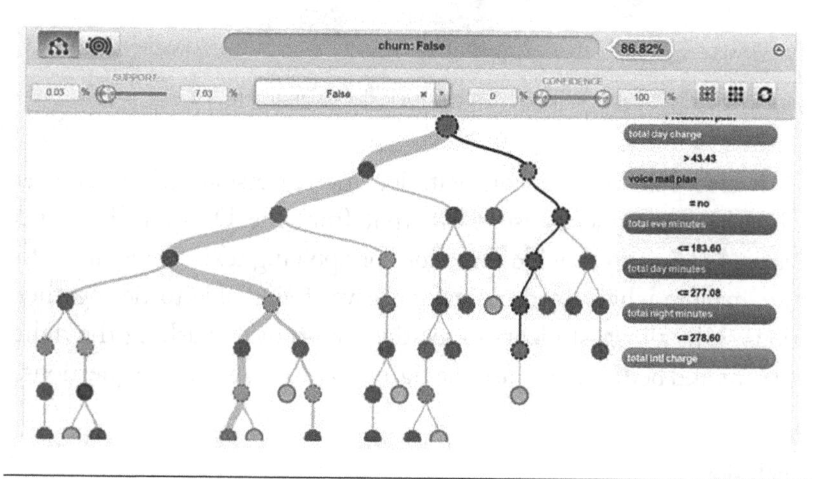

Figure 9.13 Visualization of a decision tree when the Customer Churn is predicted to be false.

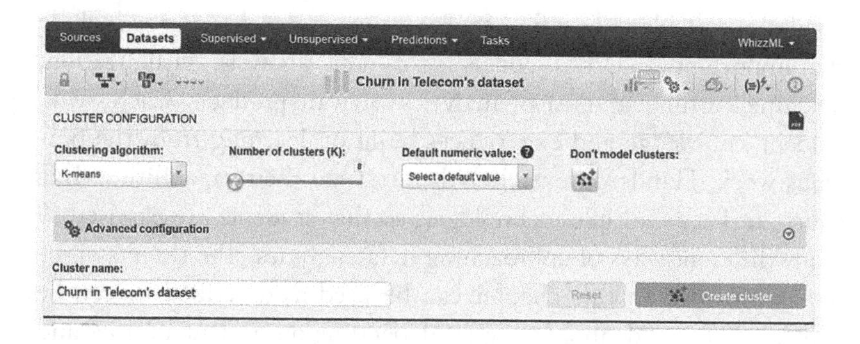

Figure 9.14 Configuration screen for clusters.

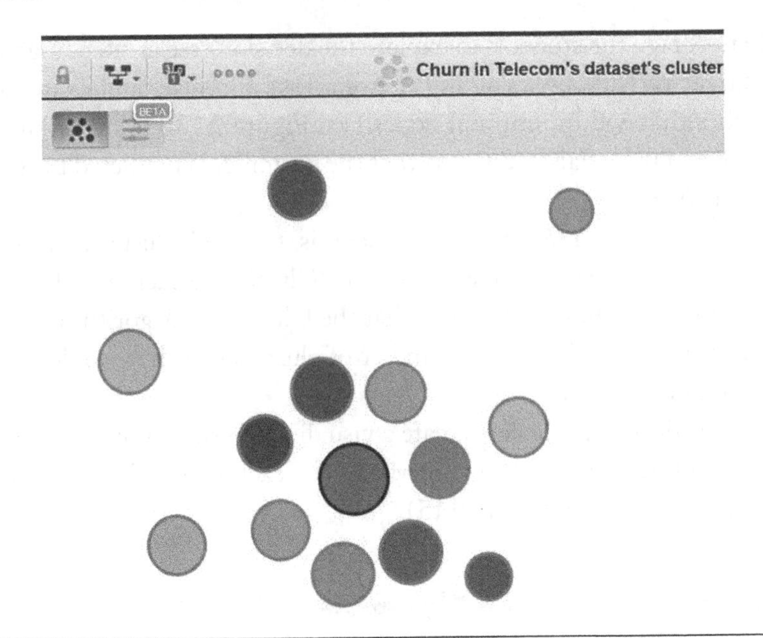

Figure 9.15 Visualization of the clusters created in the assignment ($K = 14$).

It is also possible to incorporate logistic regression assignments as well as having students use the output from the Decision Tree and Cluster Analysis to provide direction for applying text data analysis to future explore why customers stay or leave. Being able to tie together various skills allows students to see the importance of asking the right questions and being able to use the right tools to answer these questions.

Conclusion

This chapter has described an introductory course in Business Analytics suitable to be a first course for programs desiring to include the important analytics concepts of data analysis, visualization, machine learning, and use of analytics software products. Each week, student knowledge and experiences build on learning from the previous week. Hands-on exercises demonstrate their capabilities. This course utilizes multiple technologies so that students are exposed to many different ways of approaching data analytics. The three assignments provided in this chapter can be used as described or can be expanded and made more complex depending on student abilities and time allowed for the assignments.

Appendix 9A: Weekly Course Plan

Week 1: Value of Analytics and Decisions
- Definition of business analytics
- Evolution
- Value of business analytics
- Data-driven decisions
- Decision biases
- Excel Pivot table exercises
- Chapter 1: The Value of Business Analytics

Week 2: Analytics in Business and Approaches
- Analytics maturity models
- Introduction of analytics:
  - Descriptive
  - Predictive
  - Prescriptive
  - Text
- The steps of CRISP-DM
- How to approach data analytic problems
- SAP with Microsoft Excel
- Chapter 2: Producing Insights from Information through Analytics

Week 3: Dashboards and Visualizations
- Dashboards
- Examples of effective dashboards
- Types of indicators on dashboards
- Visualizations
  - What "not" to do
  - Best practices
- Lots of examples of visualizations: how do the students interpret?
- Build a dashboard in Tableau
- Chapter 3: Executive/Performance Dashboards

Week 4: Data Mining and Warehouses
- Data mining: definitions
- Data warehouses

- Data mining techniques: prediction, classification, clustering, association
- Dashboard with Tableau
- Chapter 4: Data Mining: Helping to Make Sense of Big Data

Week 5: Clusters and Big Data

- Clusters
    - Definition
    - Applications
- Big data
    - Definitions
    - Applications of big data in industry
- Unstructured data
- Background about R program
- SAP Predictive Analytics clusters or Tableau clusters
- Chapter 5: Big Data Analytics for Business Intelligence

Week 6: Predictive Analytics

- Classification
    - Definition
    - Applications
- Supervised learning
- Linear regression
    - Definition
    - Applications
- Logistic regression
    - Definition
    - Applications
- SAP Predictive Analytics—Automated Analytics (Classification)
- Lynda.com classification and Big data, Analytics, and the future of marketing and sales (McKinsey)
- TED talk: How we're Teaching Computers to Understand Pictures www.ted.com/talks/fei_fei_li_how_we_re_teaching_computers_to_understand_pictures

Week 7: Data Structures

- Relational databases
- Entity relationship diagram
- Examples of diagram

- Structured query language
- Data warehouses
- Star schema
- Snowflake schema
- Practice SQL language in quiz

Week 8: Advance Data Structures
- Hadoop
- In-memory processing
- SAP HANA Sales Monitor

Week 9: Text Analytics
- Text analytics
- Where does text come from?
- Data preparation
- Analytics used in text analytics:
  - Clusters
  - Predictive analytics
  - Logistic regression
  - Single-value decompositions
  - Decision trees
  - Visualizations of correlations and frequency (word cloud)
- Applications of text analytics
- Text analytics group project: TripAdvisor restaurant reviews
- Chapter 6: Text Mining Fundamentals

Week 10: Machine Learning
- Machine learning
  - Definition
  - Applications
- Neural networks
  - Definition
  - Applications
- BigML, Inc.—Telecom churn database
- Articles about machine learning

Week 11: Mapping
- History of mapping
- Information on maps

- Elements of a map
- Baselayers
- Examples of maps with analytical data
- SAP Lumira mapping and ESRI
- Read article in mapping journal, and write a summary to be presented in class

Week 12: Social Media

- Social media by the numbers
- History of each social media
- Application to business of various social media
- Algorithms using social media data
- Chapter 8: Measuring Success in Social Media: An Information Strategy in a Data Obese World

Week 13: Legal and Ethical Implications

- Privacy concerns in the form of questions
- Definition of privacy
- Data mining by government agencies
    - The United States
    - Europe
    - Other countries
- Regulations in the United States (FCRA, HIPAA, FERPA)
- Definition of ethics and dilemmas
- Current issues in industry
- Chapter 9: The Legal and Privacy Implications of Data Mining

Week 14: Industry and Next Steps

- Examples of job titles and descriptions
- Continuing education opportunities
- Free resources
- Chapter 10: Epilogue: Parting Thoughts about Business Analytics

Week 15: Final Exam

10

Building a Ranked Data Analytics Program

VIRGINIA M. MIORI,
NICOLLE T. CLEMENTS,
AND KATHLEEN CAMPBELL-GARWOOD

Saint Joseph's University

Contents

Introduction

Although the name changes with alarming frequency, the field of quantitative study known as data analytics is founded in well-established techniques, tools, and models. It began through the cross section of operations research/management science, information systems, and statistics. This broad collection of techniques complicates the process of creating and deploying new programs in the area of data science and data analytics.

According to Davenport and Patil (2012), a data scientist is a high-ranking professional with the training and curiosity to make discoveries in the world of big data. The term data scientist gives a name to these "new age" statisticians, operations research analysts, programmers, and problem-solvers who have always existed, and it makes their role in corporate America new and exciting. However, preparing masses of data scientists to be operational in the workforce has partially fallen into the hands of Academia. Smart and driven employees are seeking to gain continued expertise across a diverse range of fields in order to be the "data scientist" in a variety of industries. A data scientist is not one position. It is a problem solver with both technological and analytical skills who can identify rich data sources, join them with other, potentially incomplete data sources, and clean the resulting set (Davenport and Patil, 2012). Therein, the skills needed to produce educated active data scientists are embedded in known methodologies like programing and statistical analysis but with an ever-changing landscape as new programs and problems arise. In order to facilitate the best learning situation, master's programs need to be targeted, challenging, and nimble.

As more programs in business and data analytics have been introduced, they have been housed in various schools within universities. Offering data analytics programs requires each department or school to reflect its own identify and mission, within the broad definition of business intelligence and analytics (BIA).

The mission statement for Saint Joseph's University (SJU) affirms that "Saint Joseph's provides a rigorous, student-centered education rooted in the liberal arts. We prepare students for personal excellence, professional success and engaged citizenship." (Mission Statement, n.d.) The mission statement for the Erivan K. Haub School of Business (HSB) within SJU asserts that we "seek excellence in business education that offers *breadth* in terms of broad-based coverage of business concepts and skills, *depth* through focus on specific industries and professions, and *wholeness* via education of men and women in service to others..." The core values of the HSB stress the development of innovative niche programs: "...the HSB has been entrepreneurial in its approach to targeting and serving the needs of key industries and strategic niches." (Graduate Business, n.d.) The Master of Science in Business Intelligence and Analytics (MSBIA) at SJU reflects the missions and the core values of the business school and the university.

In this chapter, we provide the history and creation of the MSBIA at SJU. We further discuss the continual process of adapting the programs to meet the needs of industry and the importance and implementation of prerequisite statistical skill acquisition.

Note that SJU currently offers two options for MS program delivery: online and on campus. Students are accepted into their program of choice but are offered the opportunity to take courses through both delivery methods.

MSBIA Program Creation

The Decision and System Sciences (DSS) Department at SJU was formed in 2003 and is made up of faculty with specialties in the areas of operations research, information systems, and applied statistics. The department began offering a Master of Science program in DSS in 2005. Not long after its creation, it became clear that the somewhat meaningless name and the traditional content were in need of refreshing. Enter the online Master of Science in Business Intelligence (MSBI). A committee of DSS faculty members came together to research industry needs and adapts the program to fulfill those needs.

The designation of "business intelligence" (BI) was selected because it was a recognizable industry term that provided an appropriate umbrella for the topics within the program:

> The term **Business Intelligence** (BI) refers to technologies, applications and practices for the collection, integration, analysis, and presentation of **business** information. The purpose of **Business Intelligence** is to support better **business** decision making. ("What is Business Intelligence," n.d.)

With such a broad definition, it was critical that our department begin with a cohesive sense of our own identity and our own purpose. As a department, we began by highlighting our own skills and contributions and then identifying a structure that would demonstrate the interaction of these contributions. We began with the pyramid in Figure 10.1, establishing the broad concepts and the foundational importance of information systems.

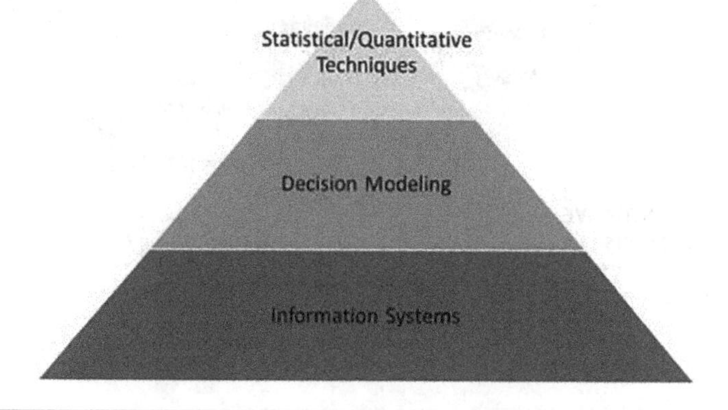

Figure 10.1 Business intelligence pyramid structure.

Knowing that we intended to span all three content areas, we moved forward in our attempt to address the vast expansion of available data. The MSBI, it was determined, would be created with a focus on modeling and techniques rather than tools. It was also agreed that the program would prepare students to become analysts and to further their progress into managerial positions for BI teams.

The next step required the exposition of topics. Ultimately, each of the levels of the pyramid would be defined and expanded to the broad instructional categories of information systems, quantitative methods, and statistics. By organizing these topics into a Venn diagram (see Figure 10.2), we were able to clearly demonstrate the overarching importance of modeling for successful implementation of BI in business (Klimberg and Miori, 2010).

The final organizational requirement was the specification of courses necessary to fulfill the MSBI. Students would be joining the program from many diverse backgrounds, many seeking the opportunity to transition to BI jobs within their own companies, while others were seeking a change in career that could offer greater promotion possibilities. Figure 10.3 shows the course breakdown. The first two courses in the program were designed to establish a common

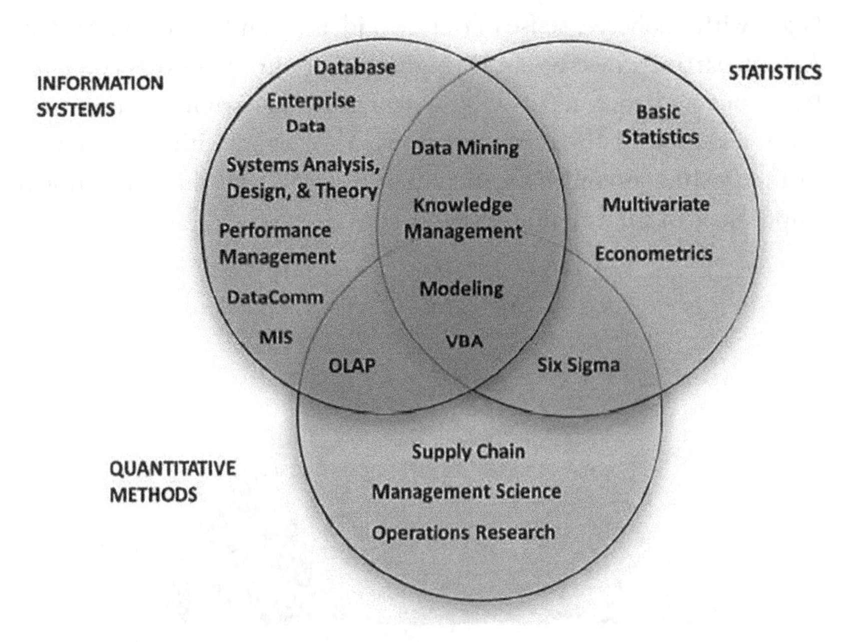

Figure 10.2 Master of Science in Business Intelligence Venn diagram.

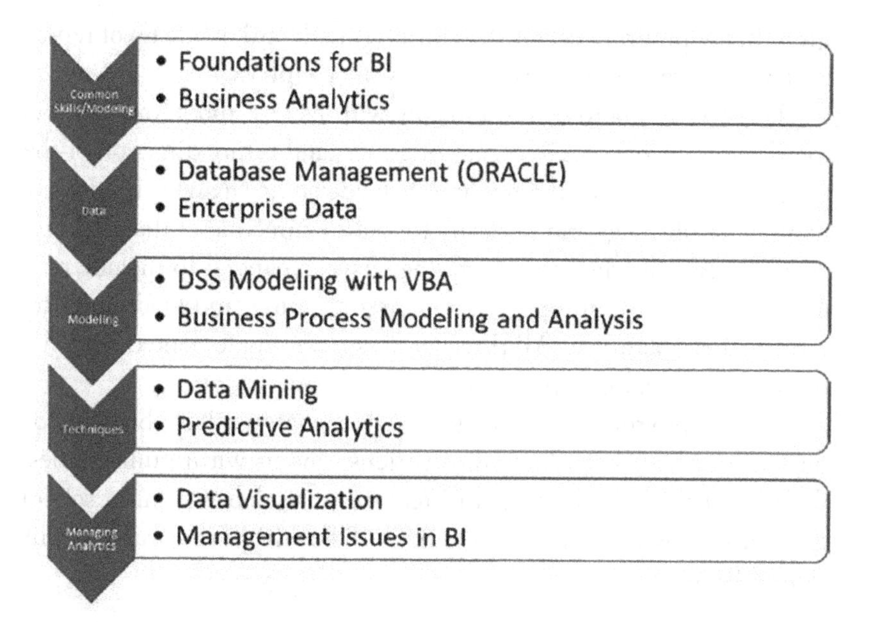

Figure 10.3 Initial course progression.

foundation among students, to introduce and/or refresh basic modeling skills, and prepare students for advanced courses. The program culminated in two courses that emphasized skills necessary for management of BI.

The online MSBI was phased in over the five semesters. Courses immediately integrated both asynchronous and synchronous content. The synchronous content was particularly critical to give our mission as a university and a business school. It was essential that students be connected to each other, to the faculty, and to the university.

Thanks in part to the 2007 publication of "Competing on Analytics: The New Science of Winning" (Davenport & Harris, 2007), demand for the MSBI expanded. Over time, terminology evolved in industry and the name of the program was adapted to more accurately reflect the prevailing job titles, it was newly titled the MSBIA.

The growth in the MSBIA graduate degrees was significant, with 50% increases in enrollment year-over-year. At the same time, competition also sharply increased. A current and innovative curriculum, with continual reflection of BIA shifts in industry; reflection of school and university missions; and adherence to school core values, are key to the success of the program. To support a vision of continued growth, the program established an advisory board, made up of representations across industries and across the country.

The advisory board meets annually with faculty and administrators to discuss industry trends, course offerings, and future direction of the program. These conversations have helped the department to identify new course offerings and to clarify the future direction of the program.

Most recently, the course offerings have expanded to include new courses in R programming, Python programming, and Hadoop using the Cloudera platform. All three courses have supplemented both the online and on-campus programs.

As more programs in business and data analytics have been introduced, they have been housed in various schools within universities. Offering data analytics programs requires each department or school to determine what their own identity will be within the broad definition of BIA.

MSBIA Program Expansion

Overall, the MSBIA curriculum has always taken a holistic approach to build the students' decision-making skill sets in a sequential manner. First, they needed to understand the data within the context of the business problem. Second, they needed to understand enough about the information system and analytics to be able to know the right tools to use and apply. Third, they needed to be able to implement the tools and methods before finally communicating their results effectively. In addition to these fundamental skill sets, it has always been very important for students to be well rounded and curious. As noted before, the BIA market is becoming competitive and one can be left behind if not working hard to stay current. Even if we teach students most of the cutting-edge tools currently used in industry, we won't be able to predict the tools they will need in 10+ years. Thus, the MSBIA curriculum has always promoted intellectual curiosity and encouraged students to prepare not only for the industry of today, but also the industry of the future.

In a recent Harvard Business Review webinar, Tom Davenport explained three previous eras of analytics, beginning in the mid-1970s, and how the industry is entering a new era, Analytics 4.0 (see Figure 10.4). This era includes new skills such as automating tasks and

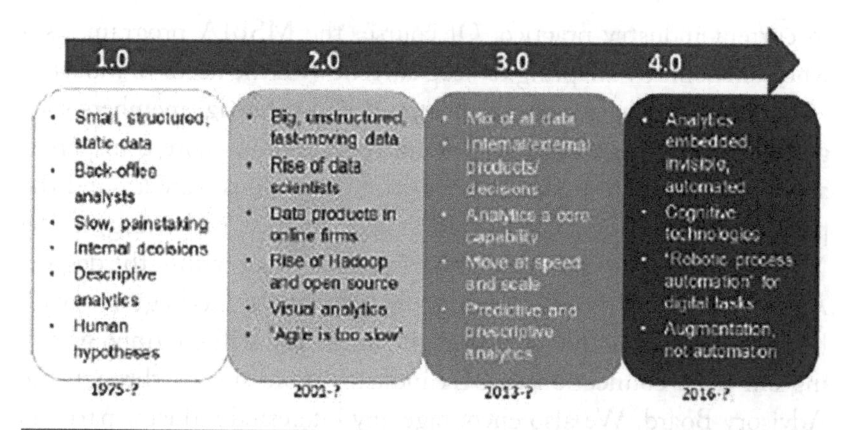

Figure 10.4 The Eras of Analytics taken from Davenport's Harvard Business Review (2016).

decisions. "Automation requires advanced data management and processing capabilities requiring use of tools such as Hadoop and Spark and a mix of proprietary and open source software." When thinking of expanding our MSBIA program, we pay close attention to the advice of industry pioneers, such as Davenport. In the MSBIA program, we have already made strides to keep pace with the idea of Analytics 4.0 by integrating Hadoop, R, and Python into our curriculum.

When looking for future directions for our MSBIA program, we often turn to our Advisory Board for advice and new ideas to stay ahead of industry demands. Through this Advisory Board, we partner with large companies in pharmaceuticals, Big Four accounting firms, telecommunications, as well as smaller-scaled local companies that are looking to hire our students. The Advisory Board meets approximately once per year with the faculty of DSS, program administrators, and Deans of the Business School. Discussions often include changes, additions, or deletions of course content, internship and employment opportunities, ideas for case studies or student projects, and strategies to market the MSBIA program to new cohorts of students. This arrangement benefits the companies, students, and faculty alike. Employers get access to cutting-edge methodology, academic experts, and bright students, as well as the chance to shape the skills they will receive with future employees. Students get exposed to internships and job opportunities after graduation. They also gain experience solving actual problems dealing with real and messy data. As faculty, we are able to keep our research and teaching connected to current industry practice. Of course, the MSBIA program, as a whole, benefits by adapting to keep current with the needs of industry.

For example, in a recent Advisory Board meeting, members suggested name changes for a few of our courses to be more transparent about the topics covered. Other fruitful discussions were around the possibility of adding a concentration (e.g., finance, marketing) in the MSBIA program to give students a second skill set within the degree. After this advisory board meeting, the DSS faculty has begun to work hard to implement the ideas brought forth. We will continue fostering our deep connection to BIA industry professionals through the Advisory Board. We also encourage any interested industry partners to teach specialized courses in the MSBIA program. This allows students to see real-world applications of textbook material, taught from an industry-professional instructor. These close associations will continue to keep our program current with industry demands.

Another way we reconfirm the MSBIA program changes is by deploying surveys to our alumni. We regularly survey recent alumni, targeting those who are working in the BIA industry, and ask pointed questions about the ways in which the MSBIA curriculum has prepared them for, or assisted in advancement in their job. As is typical with surveys, the response rate is mediocre, but the alumni survey still provides useful insights about courses and topics covered in our curriculum. The information provided by our alumni are combined with the suggestions from the Advisory Board and used to make strategic decisions about the MSBIA program.

An innovative way to keep our program flexible to curriculum changes is the creation of "tracks" within the MSBIA program. These tracks are a sequence of courses designed by the faculty to give guidance to students on the courses that fit together to form well-rounded skill sets. Currently, we have three tracks: Traditional MSBIA track, Data Analytics track, and Programming track. The DSS faculty, along with input from the Advisory Board, are establishing a fourth track in Cyber Analytics. Tracks allow our program to remain flexible, meeting student needs beyond the generalized degree, and supporting the changing demands of industry. Tracks are also made available to MBA students, wishing to concentrate in the area of BIA.

In addition to creating tracks, we are launching postbaccalaureate certificate programs. These four-course certificate programs are designed to help degreed professionals advance their current skills obtained from generalized programs in BIA without committing to the full ten-course MSBIA program. The initial certificate programs draw heavily on two of the tracks within our program: data analytics and cyber analytics.

One way that we encourage students to be prepared for future industry changes is gaining familiarity with Massive Open Online Courses (MOOCs) such as Coursera and edX. MOOCs are free online courses available for anyone to enroll. MOOCs allow anyone with desire and diligence to improve his or her skill sets. It has never been easier to acquire new skills. For example, as of December 2017, a search for "data" in the Coursera course catalog returns 871 courses and specializations at varying levels of expertise. In the MSBIA program at Saint Joseph's University, we have a fundamental focus on building our students' skill sets to understand the business problem, properly choose and implement the methodology, and communicate the results. A secondary focus is to stimulate their intellectual curiosity so that students are not only prepared for the industry of today through the MSBIA program but are also able to learn and adapt to future industry changes after they've graduated.

Statistical Skill Acquisition

As mentioned in the MSBIA history, a graduate program producing analysts capable of fluidly engaging in analytics roles is daunting. Students who wish to move in this direction come with a wide variety of backgrounds ranging from computer programming, corporate or personal finance, accounting, healthcare, military, and engineering to name a few. Each student brings a different skillset and work experience. In particular, students have varying statistical skills.

Initiating a program with a basic statistics course that fully encompasses traditional undergraduate topics (basic probability, hypothesis testing, and regression) is unrealistic given the variation in needs among students. It would challenge the patience of strong students while pressuring students who lack a solid base. In addition, it would add an 11th (prerequisite) course to the MS degree.

Rather than adding an additional course to the master's program, the department chose to provide a less expensive, self-paced learning tool for all students. Students are encouraged to begin using this tool prior to their first foundational course, but completion of the proficiency is required and counted toward the grade in that first course. This allows each student to work through the necessary topics at his or her own pace, to achieve a baseline understanding required for success in upper level courses.

The ALEKS program, offered by McGraw Hill, was selected. ALEKS is a leader in the creation of Web-based, artificially intelligent, educational software ("About Us" n.d. web accessed: Aleks, 2017). After carefully considering the 161 possible topics in business statistics, 137 topics were selected as necessary for all students in the MSBIA to master. Each student completes an initial assessment with a maximum of 25 questions. An artificially intelligent engine analyzes correct and partially correct answers to establish their degree of mastery of the 137 required topics. A pie visualization is then personalized for each student, with slices of the pie being shaded to indicate proportion of achieved knowledge acquisition and remaining topics requiring study. Advanced students who show mastery in more than 88% of the topics are exempt from completion of the ALEKS module.

It became clear very quickly that the materials available within the ALEKS tool were insufficient in completely supporting student needs. To mitigate this issue, a set of video tutorials, addressing each of the 137 topics, was created by a very enterprising faculty member. These tutorials are invaluable in supporting students in achieving statistics proficiency. While demanding, the ALEKS program allows all students to work within their own comfort level and to be challenged meaningfully.

Unanimous department endorsement of the ALEKS proficiency tool provides clear evidence that the statistics proficiency requirement has prepared effectively students for courses at all levels.

Concluding Comments

As with any industry segment, success can only be achieved when new initiatives fit within and further the organization's mission and values. Establishment of our MSBIA programs began with SJU and HSB missions and HSB core values. Success derived from our conscious choice to carry these principles through the structured process of course identification, development, launch, and administration. Attention to identity and purpose are never wasted and often result in unexpected gains.

References

About Us (n.d.). ALEKS. Retrieved from https://www.aleks.com/about_us

Davenport, T. H., & Harris, J. G. (2007). *Competing on Analytics: The New Science of Winning*. Boston, MA: Harvard Business School Press.

Davenport, T. H., & Patil, D. J. (2012). Data scientist: The sexiest job of the 21st century, *Harvard Business Review, 90*(10), 70–76.

Graduate Business (n.d.). Saint Joseph's University. Retrieved from http://www.sju.edu/int/academics/hsb/grad/about/mission.html

Klimberg, R. K., & Miori, V. M. (2010, October). Back in Business, *OR/MS Today*.

Mission Statement (n.d.). Saint Joseph's University. Retrieved from https://www.sju.edu/mission-statement

What is Business Intelligence (n.d.). OLAP.com. Retrieved from http://olap.com/learn-bi-olap/olap-bi-definitions/business-intelligence/

Index

N

O

P

Q

R

S